Scott Eastham is currently Senior Lecturer in the School of English & Media Studies at Massey University, Palmerston North, New Zealand. He is the author of *Paradise & Ezra Pound, Nucleus, The Media Matrix, The Radix, EyeOpeners*, and most recently, *The Way of the Maker* — an edited volume of essays on art and spirituality by Canadian artist Eric Wesselow.

Scott Eastham

BiOTECH
TiME-BOMB

How genetic engineering could
irreversibly change our world

RSVP
PUBLISHING

For Peter Horsley

First published in New Zealand in 2003 by
RSVP Publishing Company Limited
PO Box 47 166 Ponsonby
Auckland
www.rsvp-publishing.co.nz

Book Design by Stephen Picard

Typeset in Times

Printed in New Zealand by Bellprint Limited.

ISBN: 0-9582182-2-6

CONTENTS

Worldviews in Collision —
The Challenge of Genetic Engineering

In the world's first such public inquiry, New Zealand's Royal Commission on Genetic Modification accepted and reviewed nearly 11,000 submissions. Ninety-two percent of those commenting on the possible use or abuses of this new technology opposed field trials or commercial release of genetically modified organisms. Yet, with a flourish of precautionary language, the Commission's report (August 2001) proposed to allow both. Its recommendations now form the basis of government policy on genetic engineering in New Zealand.

Similarly, in late 2002, the Environmental Risk Management Authority (ERMA), charged with overseeing such trials and releases, permitted transgenic calves (incorporating human genetic material so as to produce human proteins in their milk) to be reared on a fenced-off farm in the Waikato. In this case, ERMA had received 856 submissions from individuals, local iwi, and other organisations opposed to the field trials, and only seven (less than one percent) in favour. Yet approval was granted, opening the door for all sorts of further human/animal chimeras.

Granted that neither the Royal Commission nor ERMA were merely taking polls for or against genetic engineering in New Zealand, the results of their extensive deliberations seem bizarrely out of balance with public opinion. Why should the 'expertise' of a few scientists and corporate lawyers (all with obvious financial interests) carry so much more weight, or be deemed so very much more 'objective' than any other testimony? We cannot ask the Royal Commission; it has completed its work and disbanded. But unless we attribute bias or venality to the hard-working commissioners or to ERMA, we need to seek other reasons why such well-intentioned deliberative bodies routinely dismiss the overwhelming proportion of testimony submitted to them.

Part of the problem may lie in the formal structure of government 'hearings' altogether. The Royal Commission, for instance, was explicitly set up in adversarial fashion as a mock courtroom, complete with witnesses and lawyers from the pro and anti-GE contingents to examine them. As a medium of communication, such a forum can only be considered a fairly clunky apparatus. It is about as far from a normal human conversation as it is possible to get; 'conversation' is what people do in the hallways before or after they make their formal 'statements' for the record. A brief 'Question and Answer' period may be allotted for each witness, but this limited give-and-take rarely changes any minds. The system itself operates mainly as a one-way channel of communication, like those old single-sideband ham radios they used to use on remote sheep stations, where you first set up the machine to receive (any and all messages from the public welcome), then flip a switch to transmit (the ruling or conclusion, broadcast on every frequency), and finally just shut off the whole expensive, finicky contraption before it burns out or breaks down.

Another limitation lies in the peculiar character of 'testimony' itself, a once-and-for-all statement which precludes dialogue. The very act of bearing witness — "This is what I saw." "These are the results of my experiment." "That is my belief." — stops any possible dialogue dead in its tracks. What can you say to that? Bearing witness is in this sense an ultimate act; you cannot get beyond it. Although conflicting testimonies might 'raise questions' for the panel, the witnesses themselves cannot speak directly to one another on the record, and are of course never invited to sort out their differences. Ironically enough, the prevailing regime of formal debates and public hearings — which only encourages opposing factions to maintain their customary fighting stances — may well have guaranteed that genuine dialogue on this issue has yet to get underway in New Zealand.

Each witness will, moreover, speak to these panels in their own (professional) 'dialect' or (personal) ideolect. Scientists, ecologists and technicians, all very knowledgeable about particular genetic techniques, speak — of necessity — in the highly specialised jargons appropriate to their disciplines. The lawyers and bureaucrats predictably tend to utilise their most obtuse legalese and bureaucratese vocabularies for such occasions; sometimes they even appear to understand one another. Corporate interests offer upbeat risk/benefit analyses solely within their preferred — and deliberately restricted — financial frames of reference. Maori elders speak eloquently of their deepest cultural values, but to do so must stick religiously to their traditional terminology. And, oh yes, plenty of so-called 'ordinary' New Zealanders — farmers, fishers, local business people, clergy, teachers, greengrocers, ecofeminists, sick children with genetic maladies — are also keen to have their say, each in their 'extraordinary' way. Although their testimonies are presented in English, and are all duly transcribed for the record, it is plain that not everybody is really speaking the same 'language'. These divergent ways of speaking are often just plain incompatible, for the simple reason that they stem from and attempt to articulate very different, and often very deeply rooted worldviews. At the end of the day, the 'official' recommendations generally confirm that only certain sorts of technical language — scientific, legal or financial jargons — are taken seriously by government panels. To many of the ordinary people who participate in such 'hearings', usually at their own expense and because they care deeply about the issue, it may seem as if they were never 'heard' at all.

It is my contention here that far more may be said, indeed must be said, in ordinary language about these new technologies than can ever be said in any specialised technical jargon. In fact, technical thinking is incapable of thinking about technology (or anything else) as a whole. Only the culture at large — with its history, arts, literature and philosophy, as well as its critical traditions — can provide an overview of these techniques which does not obscure their origins in human visions, ambitions, inventions, decisions, blindspots, etc., or else blur in a flurry of fuzzy jargon their eventual impact on human life. The common language we need, and which in a sense we do already share, is part and parcel of the entire tough, sinewy fibre of culture — neither a superstructure nor an infrastructure, but the very metabolism of human awareness circulating between the individual and the collective dimensions of our experience. If a culture produces a certain technology, it is to that culture we must turn if we are ever to understand it.

So it is time to open up the strenuous and yet unavoidable dialogue on genetic engineering, an issue which may soon prove to be ultimate and decisive for humankind, as well as for the many species with whom we share the biosphere. Such a dialogue is not just a debate on television, nor is it subject to legal or scientific 'proofs', nor can it be fully expressed in the dialectical arena of political protest. It needs to engage worldviews which seem implacably opposed to one another, and which have so far met merely to clash at the level of politics and ideology. The new techniques for manipulating life stem directly from the modern Western scientific mentality, and bear the imprint of the long history of Western culture in the very urge to control both the social and the natural worlds. Left to its own devices, Western culture simply cannot stop itself from adopting techniques which promise this kind of unlimited power, though they may threaten irreversible damage to nature and culture alike. Only an intercultural perspective, grounded in an openness to insights from other worldviews, may offer any hope of a corrective. Human cultures one and all are today challenged to come to grips not only with our newfound powers to manipulate life, but with our long and sorry history of failing to do so wisely.

When Worldviews Collide

I wish first of all only to briefly introduce several possible points of departure for that dialogue, centering on the ethical questions raised by these new technologies:

• *Science does not have all the answers.* Responses to biotechnology are generally deemed to be of two types: scientific and unscientific. I would maintain that both are valid, but they are certainly not commensurate. The scientific responses, whether positive or negative, at least share the worldview of modern science... an objectifiable world of matter, mass and motion, mathematical proofs and mechanisms which may be empirically tested, and which will yield certain sorts of information considered reliable within that scientific worldview. From this quarter, responses to biotechnology tend to focus on the possible risks or benefits of applying these techniques, on the lack or the extent of our knowledge of natural organisms and processes, on the viability of alternative techniques, and so forth — in short, on what is good science or sloppy science or bad science, given our current knowledge. But the scientific worldview is not itself in question here; everybody shares the same assumptions, and the argument has the character of a rational debate. You may already be aware that there is no unanimity about these techniques on scientific grounds alone. But there is another sense in which science does not have all the answers... Reason can only resolve a dispute if everybody agrees on the rules and, indeed, agrees to play the same game. Yet biotechnology raises issues which amount to *a clash of cosmologies*, a collision of worldviews. 'Unscientific' objections to these techniques are too often ruled out of court at the outset because they do not share the assumptions of scientific method. Yet we know very well that for all its pretensions to universal truth and factuality, science itself is a system of belief — in its methods, in its presuppositions, and indeed, in its sources of funding (which in turn 'trust' those methods and presuppositions) — quite as much as any other human endeavour. To put a sharp point on it, people do not support or condemn genetic engineering because they have more or less information. They support or condemn it because they believe in it, or they don't. All the information in the world only adds to that bedrock of conviction. This is why dialogue is paramount here, and not merely legal wrangling or debate on grounds laid down by scientists, but constructive dialogue where each partner assimilates and builds upon at least some of the insights of the other. Only the other can show you what you take for granted, only another worldview can reveal the limits of your own. Only the other can question what you you consider unquestionable.

• *The question of evil.* Sister Miriam MacGillis of the renowned Genesis Farm in the USA, a strong proponent of organic farming as well as a religious woman, says it in so many words: "Genetic engineering is evil." How do we evaluate such a statement, which so obviously stems from a different worldview than that of modern science? Who decides if this is good or evil? What, indeed, is evil? Would we know it if we saw it, or if we were doing it? And beyond personal morality, how does evil get into systems, into technologies, into bureaucracies, into institutions? These are questions which science and engineering do not pretend to answer, and which all too often tend to be dismissed as some sort of primitive superstition, despite the 20th Century's massive experience of systemic evil, in the concentration camps, for example, in the Manhattan Project's nuclear weapons, in transnational corporations plundering irreplaceable natural resources, in global financial institutions squeezing the economies of the Third World, and so forth. No, we'd rather not talk about evil... we would rather say something is a mistake or an error, or maybe not cost-effective, or else ill-informed or misdirected. But evil? Where are the experts on evil? What do we know about evil? One thing we know for sure is that the worldview of modern science alone is incapable of dealing with it.

• *Focus and blindness*. Molecular genetics is a science of bits and pieces, of means to specific ends. It is very clear about what it sees in its microscopic field, but very vague about the social and environmental context in which such techniques may be applied. Every increase in focus entails a corresponding expansion of the field of blindness. Is there an ethical response to genetic engineering *in principle*? Ethics deal with ends, and with the whole. It is concerned precisely with the big picture left out of the scientific focus. Ethics cannot be objectified; human subjects are not (yet genetically modified) objects. Ethics is not a science, nor is it a set of rules and regulations. The word *ethos* means 'way of life', character of life. The way of life promised by unbridled application of genetic engineering techniques is an artificial world or, to be exact, an 'artificious' world, a world produced by artifice. This is dangerous because life can only be planned or controlled to a point, a point beyond which it asserts its own indefatigable creativity and spontaneously produces the unexpected: wonders and marvels of course, but also atomic bombs, ozone depletion from CFCs, DDT in the food chain, etc. We know from hard experience that the side-effects of any new technology often turn out to be the main effects. We shall examine this dynamic in more detail shortly.

• *It is a very tempting technology*. We are told by the proponents of biotechnology about all it will deliver, all the people it will feed, all the cures it will effect, all the wonders it will create, and all the new goodies we'll be able to buy. These claims are structurally reducible to a single one: *"You can have anything you want."* Whatever your most pressing concern happens to be — the environment, illness, starvation, you name it — the ingenious bioengineers will cook up a solution for it (and if not right away, well, just wait for nanotechnology). This has a familiar ring to it. Where have we heard this claim before? Maybe every generation has to play with fire in its own way... Biotechnology would seem to be our current temptation, but the pattern is well-known. In fact, it's the oldest temptation in the book. It was once offered by Adam Smith's capitalism and then again by Marx's communism; it has been promised in our own time by dictatorships and democracies alike, by bureaucracies and technocracies of every political stripe, by religious reformers and by wild-eyed zealots. Not so long ago, nuclear energy was going to solve all the world's problems. In the 19th Century, *laissez-faire* capitalism was the solution. And once upon a time, conversion to Christianity would solve, or at least salve, all of humankind's woes. Indeed, this very temptation was what the Devil offered Christ himself in the desert; and it was among the grand illusions with which Mara tried to tempt the Buddha under the Bodhi tree. "Over all this, I will give you dominion..." At least these two worthies saw through the ruse. Has our modern scientific mentality 'evolved' so far that we no longer even recognise this temptation for what it is?

• *The myth of progress.* The most conservative and traditional imaginative projection of Western culture over the past two and a half millennia has been to project (and indefinitely postpone) salvation to the 'next' world. It's usually a bright and shining excuse for not facing this present world where people actually live and die, suffer and hope. Biotechnology is the latest 'evolution' of this hope (or should I say blind faith?) invested in evolutionary 'progress'. After millennia, we have learned from painful experience to look a little askance at this myth of progress. Maybe we've seen too many science fiction films to trust unquestioningly in the good faith and 'disinterested' research of the scientists anymore (not to mention the governments and corporations who fund them). More to the point, we touch in the myth of progress the very basis of Western cosmology. This is the worldview of a people *on their way* through time, space and history to the 'next' world, the promised land, the heaven above, the utopia, the classless society, the good life, the great society, or lately the information society and the coming 'biotech century'. The history of religions shows us that it is in the creation of order, *any* order, that evil (or the possibility of *dis*-order) arises: The world was created, and along with it came the possibility *that it might be otherwise*. So we decide to fix it... and many of the great achievements, as well as the Frankenstein monsters spawned by Western civilisation, derive from this tremendous (and sometimes terrifying) freedom we have to explore what is possible. Certainly there is plenty of good science to be done here, and much to be learned. But in today's competitive, commercial climate (the social offshoot of the same mindset), research and development are pronounced as if they were a single word, or at least a single unstoppable process. That's progress for you!

• *Wrong paradigm?* We are assured by many practitioners of these techniques that genetic engineering will confer a survival advantage upon the human species, i.e., help our species survive. If we accept this, we have assumed that survival in the evolutionary game amounts to winning a competitive struggle with other forms of life, and that by putting them at our service we will be the winners. Slave-holding societies might have been comfortable with the idea of corporations owning patents on living beings, but are we today ready to play Masters of the Universe in earnest? What if this Neo-Darwinian paradigm turns out to be wrong? Certainly many other cultures, particularly indigenous place-centred cultures, including Maori here in Aotearoa, see the entire relationship of humans to the lifeworld differently. What if they are right and the key to survival lies in co-operation, harmony with other lifeforms, mutual sustenance and support? Then by presuming ourselves the lords and masters of all creation, we might well be writing our own death warrant as a species. Are we willing to submit to proof by disaster? Nothing is more blinding than a bright light. Perhaps a fundamental paradigm shift is required of bio-science... before the effect of its manipulations become irreversible.

• *The Great Experiment.* Let us accept for a moment the claim so often made that the momentum of this technology is irresistible, that the entire planet is becoming one vast biotech lab, with all biosystems as the testing ground, and all human beings eventually guinea pigs for the great experiment. Now I submit to you that, besides flagrantly violating the juridical principle of informed consent, this is not 'sound science' at all, but downright sloppy experimental method. There is no 'control' group, no bioregion and no population NOT subjected to these techniques of genetic manipulation. In this case, rather than choosing conformity, maybe

New Zealand should choose to be the 'control'. Might we not serve the world community and indeed the interests of 'sound science' best by preserving on our islands at least this one genetically unaltered environment, so that the 'results' of the great experiment may one day be assessed in proper scientific manner? Otherwise, who knows? Who CAN know, in the restricted scientific sense of knowing by experimental method, what the results of re-engineering life on this planet will be? Thousands of labs around the world are churning out genetically modified organisms — from mighty mice with huge muscles to tobacco that glows in the dark to citrus vesicles that grow entirely in the lab, without soil or farm, on tissue cultures in vats of nutrients. We may not be able to turn this tide, but when the tide does turn — as it will when the one-in-a-thousand or one-in-a-million irreversible accident occurs and people begin to have second thoughts — it would be nice to find one corner of the world preserved from this wildfire genetic time-bomb. This place isn't Shangri-La, but it could turn out to be Noah's new Ark.

• *Interdependence.* From the point of view of human ethics, what we have to learn from these new techniques, either in the lab where they belong or (heaven forbid!) let loose in the field, amounts to a revision of the current paradigm and a rediscovery of the law of *interdependence* which already prevails between and among all forms of life on this planet. People these days have trouble seeing this interdependent web of life; our technological civilisation has fostered the illusion that we are somewhat independent of nature, insulated from heat and cold, day and night, or the seasonal rhythms. It might be simpler to see if we transferred the dilemma from the biological realm to the social or economic spheres. Let's say I have spliced up a little bacterium that eats silicon (perfectly feasible), and so threatens to gobble up the central processing units of all the computers in the world... I suspect that the powers that be would very quickly impose limits on the field testing of my invention. The fact is, we're smack in the midst of a goldrush in biotechnology, and it's difficult to stop a stampede. But let's suppose we could cut out all the intermediate products and simply engineer a money tree which produced, say, nice crisp hundred dollar bills instead of leaves.

We would be very quick to invoke the interdependence of global markets and restrict the field testing of money trees. Unfortunately, nobody has yet located a gene for greed, so it is difficult to eradicate the urge to get these things out of the lab and into the marketplace as fast as possible so that they will do what they are mainly intended to do, namely, make bundles of money for those who own the patents on them. And then, unfortunately, we will have to learn our lesson of interdependence the hard way, when the inevitable miscalculation appears and there suddenly are no more monarch butterflies, or a disease worse than AIDS appears, or the food chain begins to die out from under us. A hundred and fifty years ago, New Zealanders began burning off the bush, to the effect that 70 percent of the native growth of these islands is now gone. Do we want to be the generation that replaces nature altogether? Once upon a time, introducing possums from across the Tasman looked like a real easy money-spinner. New Zealand includes a sampling of all the bioregions and climatic zones of the planet: We're a miniature Earth. We are very particular about what bugs and barnacles and toxins we allow into the country, and rightly so; the ecological balance could easily be upset once and for all. So why subvert this policy from within by engineering our own 'alien' organisms?

The Golden Rule of Bioethics

I propose taking another tack. Following the anthropocentric emphasis of Western ethics, the Golden Rule has always been formulated to underscore reciprocity, mutuality and interdependence between people: "Do unto others as you would have others do unto you." Even Darwin (in *Descent of Man*) considered moral evolution to this level to be our distinctive achievement as a species; he coined the phrase "mutual aid" which Kropotkin later used so effectively to counter the imperialistic pretensions of social Darwinism. Our technologies, however, and most emphatically genetic engineering, extend the human 'domain' (if not the Old Testament 'dominion') across species boundaries until we have today the power to control the evolution of entire species. It is time to similarly extend the reach — the 'response-ability' and mutuality — of human moral evolution.

The Golden Rule for the biotech age ought to read: *Whatever you do to the least of these creatures, you do to yourself.* And this for at least three reasons:

1/ The same technologies can be applied directly to human beings — if you can clone a sheep, you can clone a man or a woman. Indeed, such efforts are already underway, outside any moral or legal frameworks.

2/ It's a good bet that one day the chickens will come home to roost — in an interdependent biosphere, we cannot insulate ourselves from any mishaps, which will in some way affect every generation of humans to come, *forever.*

3/ And, more subtly but perhaps more importantly, our understanding of ourselves must today be transformed. We must rediscover our own identity as something broader and deeper than just the isolated individual of the modern period (the consumer of commodities, etc.). We must learn, or relearn from traditional societies, to see the human person as a nexus in the net of biological relationships which in fact constitute us. When the dusky sparrow became extinct, I was diminished; when the monarch butterfly is reportedly threatened by genetically engineered corn, it injures me and, frankly, I take it personally... In short, our new technologies require a comparable growth in moral evolution. The 'other as self' of the Golden Rule, or of Kant's categorical imperative — to treat people always as ends, never as means — are without doubt advances on *lex talionis*, but they still remain circumscribed to the realm of human interactions with other humans. They lack the moral imperative of harmony with the myriad of other lifeforms that sustain our lives, traditionally emphasised by all those 'other' worldviews with which science refuses to dialogue, maintaining its presumed right to dictate not only to other cultures, but to all future human generations, as to what kind of world (and unnatural nature) they are to inhabit or inherit.

Science's arrested moral development first hit home with the nuclear issue: You cannot threaten all life on Earth for any limited human goal. The two issues are linked; we must be very careful when we start messing with nuclei; both technologies implicate the viability of all life on this planet. Since Hiroshima, science plainly has had some catching up to do in the moral sphere. In civilisational terms, the cost/benefit analysis we apply to biotechnology today is a rather recent experiment. It is also an unduly restricted sort of accounting when the benefits accrue only to the greed of fictitious 'artificial persons' (limited liability corporations), while the risks to be borne by actual living beings and 'natural persons' alike are certainly unknown and literally incalculable.

An immediate application of this Golden Rule of Bioethics might be an ethical test for those wishing to market genetically altered organisms... that they first have their own legal codes altered, a simple change in the operating 'programmes' under which they do business, so that every shareholder (and all their descendants) be liable (fully, legally and financially) for any and all risks, for all time. In other words, if you're planning to do unto others, be prepared to be done unto according to this law of interdependence which already rules the biosphere in which we do in fact "live, and move, and have our being."

A particular instance of applied biotechnology may or may not be evil, according to your worldview; somatic intervention in health care, for example, is a matter of choice for the patient who must decide whether he or she is willing to bear the risk of gene therapy. But what is indubitably unethical is the imposition of a single worldview (the flat, scientific, desacralised modern one) not only on every culture in the world, but on every future generation. Some things are irreversible — murder, for instance, or a nuclear holocaust, or germline intervention (affecting the reproductive cells) in biology. In our ignorance, and our arrogance, we proclaim that our present scientific knowledge (of transgenic chimeras that do not even exist yet!, for instance) is so complete and so perfect that all cultures and all future generations of human beings must live with it (and with these monsters) for all time. If we persevere blindly mixing and matching species, the future will curse us for it; our children's children will not forgive us our transgenic trespasses.

The alternative we need to explore is of course genuine *dialogue* — with other worldviews and other cultures, of course, and maybe even with other species. I don't mean just chatting with dolphins, but allowing ourselves to learn from the natural world what each lifeform has to teach us *on its own terms*. Ecology has all too often tended over the past two decades to devolve simply into cost accounting and resource management. Instead, we need an attitude which Raimon Panikkar has called *ecosophical*, a willingness to get in touch with the wisdom (*sophia*) of the Earth (our *oikos*, our dwelling place), if we are even to survive on this planet, let alone flourish. One largely unacknowledged threat of these new technologies is that they will cut communication between humans and other lifeforms. What exactly can we expect to learn from the trees and flowers when every leaf bears a barcode, marking it as something judged to be a profitable commodity in the year 2003, or from the life-ways of the birds and the bees when sexuality itself has been replaced by 'more efficient' asexual cloning?

In his famous book on *Good and Evil*, Martin Buber defined evil as "the lie against being." What, then is this lie in the context of biotechnology? It is the partial perspective posing as the totality, the single viewpoint claiming universal validity. The only way to make sure that biotechnology is not such a lie, such a monstrous falsifying of life itself, will be for the scientists to enter into long and deep dialogue with other living traditions of meaning and value. At least the cosmologies of traditional cultures (including those of the pre-modern West) have over millennia proven themselves more or less viable and sustainable *ways of life*; the mechanistic cosmology of modern science has patently failed to do so in our own time, and nobody has ever even tried to live in the artificial 'post-modern' cosmology promised by biotechnology, which to my mind has a very long way to go before it proves itself to be a good, or true, or beautiful, in any meaningful human sense.

The single example of the Human Genome Project, perhaps the example closest to home, may suffice to sum up the dilemma we humans face today.

The Human Genome: Map is not territory

Not so very long ago, sequencing the Human Genome was hailed as the Holy Grail of biology, the Rosetta Stone of life itself, and the greatest scientific advance of all time. Just recently, however, our expectations from this research have been drastically lowered.

It turns out humans have only about 25-30,000 genes, less than half the expected number, and they don't seem to explain very much after all, at least not by themselves. Another few years, we're told, and another few billion dollars, will be required to figure out how genes interact with one another, and how the proteins they generate come to form either positive traits or debilitating illnesses.

Should we be disappointed? In 1990, DNA pioneer James Watson claimed that 'decoding' the genome would tell us all there is to know about being human. Of course he was drumming up public money for the project at the time, but when he said "Our fate is in our genes," many people believed him. Others echoed this lucrative catchcry, including the University of California's Robert Sinseimer, who declared that the genome "defines a human."

Over the past decade, accordingly, the gene became a master metaphor in our society. Beyond the mechanisms of health and disease, it has been used to 'explain' everything from genius to learning disabilities, from family breakdown to homosexuality, from social privilege to sociopathic violence, from obesity to depression. It has also been used to explain away traditional forms of guilt and responsibility: "My genes made me do it."

The myth of the all-powerful gene supports a conservative view of society, where there's not much room for humans to be reformed or transformed, and therefore, not much that can be done about poverty or inequality or injustice... except of course to re-engineer those genes. All through the 90s, 'nature' trumped 'nurture' at every turn, especially in the USA, which prides itself on pioneering these techniques. For a nation originally built around notions of freedom and mutual responsibility, the 'Land of the Free', was remarkably quick to adopt what amounts to a totally deterministic view of the world. But, back to the 'incredible shrinking genome'... What does it mean when the most extensive — and most expensive — scientific undertaking of all time nets us so very much less than promised?

On the surface, it means the scientists will have to scramble to keep their funding. They'll try to revamp the old promise of the genome, placing it just a little further over the horizon: "The biology is all in the proteins," Celera Corporation chief Craig Venter now tells us. You can almost see all the venture capitalists lining up to add some proteins to their portfolios.

But the real dilemma cuts much deeper. Who are we anyway? Wasn't the Human Genome Project supposed to tell us? Well, not really. The genome project has produced a map, an abstract description, an important 'view' but hardly the whole picture of what it is to be human. A map can guide you through a landscape, but it cannot predict what will happen to you once you get there. And in this case, the 'terrain' is a human being... like you or me, for instance. Map is not territory, and it would be dangerously shortsighted to mistake the genome map for the whole 'territory' of being human.

So where do we look now? The answer to any question depends in great part on the kind of question asked. One thing the genome project has demonstrated is that humans tend to see what they look for, and sometimes not much else. Taking another cue from Panikkar, we may discern three fundamental questions at issue here:

• Scientists ask the 'objective' question: *What is a human being?* All their methods and metaphors stem from looking at human beings as the 'objects' of inquiry. The answer would then be the sum total of all the results from all the scientific disciplines examining one or another aspect of the human 'organism'. Humans are those peculiar animals which talk, think, build, make love or make war, reproduce according to certain patterns, make a mess of their environment, and so on. But isn't something missing?

• Humanists ask the 'subjective' question: *Who am I?* After all, human beings are not only the objects of inquiry, they are also, and pre-eminently, the inquiring 'subjects'. And the human being is more than just a glitch, or an annoying source of observational error. It's 'us', it's 'me', and 'I' want to know not just the number of my genes but 'who' I am. Human beings seem to be these peculiar creatures whose understanding of themselves is part and parcel of who they are. Consider the rich moment when a child discovers her own name, her unique identity — that it is she and she alone who is called by her name, or by the words "I love you." Hardly the same as being 'identified' in a computer by your IRD number.

    • There is a third kind of question, which Western culture is only lately learning how to ask. We might call it the 'other' question, or the question of the other: *Who are you?* Our very notions of ourselves come at least partly in and through interaction with others. None of us stands totally alone; there are always 'others'. They may appear differently to each of us — but these 'others' will insist they have a right to their own ideas, just as 'you' would in their place. Today, no single viewpoint, no culture or religion or academic discipline, can claim to have the whole answer to the question of 'who we are'. As a matter of fact, not theory, 'I' would be incomplete without 'you'.

    I suspect that Watson and Crick, who added the double helix of the DNA molecule to our cultural lexicon, made a category error when they labelled it a 'code' — borrowing this terminology from Norbert Weiner's cybernetics. A code is an arbitrary system of signs, with a one-to-one correspondence between the sign and what it signifies (red means stop; green means go). Nearly half a century later, the upshot was that everybody expected a one-to-one correspondence between genes and the traits they 'encoded'; we supposed that the 'map' would match the territory in a point-for-point fashion. Instead, we are only now finding out that the relationship is much more complex. Indeed, the only adequate analogue for what goes on in the labyrinth of the DNA helix is language itself. Unlike some, I do not feel moved by the gigantic gaps in our knowledge about the gene to proclaim a 'God of the gaps' in the guise of some super-smart divine biologist with a long white beard. I am struck, however, by the 'revelation' emerging from this research that the 'word becomes flesh' quite literally in the coiled DNA of our cells.

In a very real sense, we *are* language, from our genetic substructure all the way through our entire cultural superstructure. And we don't just exchange information, we *communicate*. Language is always more than a system of arbitrary signs. It is more because it includes the speaker and the hearer as well as the message and the medium: It includes me, and it includes you, and our relationship can never be totally predicted or controlled because it is free, spontaneous, creative... in a word, *alive*.

By the same token, geneticists are now taking a second look at what they previously called 'junk DNA' — the seeming gibberish of spacer regions and repetitive DNA sequences. It may turn out that the entire genome project managed to classify only the verbs (the 'action' words which build up proteins), forgetting that even at the genetic level you also need the equivalent of nouns and pronouns, adjectives and adverbs and all the rest to pinpoint which bit does what to which, when and where and how, in order to actually say (or read!) much of anything at all.

We are indeed in the midst of discovering ourselves to be dialogical creatures — physically, as well as socially. Each of us is a unique bundle of articulated relationships. We are the physical precipitate of a genetic dialogue going on at the cellular level for millions if not billions of years, the social outgrowth of an historical dialogue going on at the level of human culture and language for millennia, and the psychological nexus of a dialogue that includes all our families, friends, ancestors and, to some extent, everybody on the planet — not just those equipped with computers and electron microscopes.

Plainly, we need to deepen and broaden that dialogue if ever we are to come to grips with our emerging technologies. Maybe we have too readily allowed the scientific narrative to monopolise the story of what it is to be human. Biotechnology critic Jeremy Rifkin has long called for just such a dialogue; his works are well worth consulting. A society that leaves such fundamental issues as human cloning or transgenic hybridisation to the 'experts' is already a technocracy, not a democracy in any meaningful sense of the word.

Our social and moral coping mechanisms seem to be very slow in keeping up with these technical advances. This 'inertia' of traditional value systems is not all bad; it means that these are weighty issues which are going to require a considerable pause for reflection. Until the modern era, after all, most traditional peoples believed they were descended from Gods or heroes. Now we blandly accept the news that there's barely any discernible genetic difference between human beings and laboratory mice.

Maybe we need to hear from these other cultures and try out some of those other ways of being human, before modern science is given the last word on the evolution of *homo sapiens*.

In America, where technology outruns reflection on it every time (think of guns, let alone of the Bomb), speculators are already racing to stake out and buy up chunks of the genome map in hopes of one day striking it rich with a cure for cancer, or aging, or whatever. It is estimated that two-thirds of the processed food in the States *already* contains genetically-modified ingredients, without any identifying labels. Before this happens here, I think those of us lucky enough to live in New Zealand ought to be talking about these issues outside of the laboratories — in the home, on the street, in the bars and churches and universities, as well as in the polls. If we do not raise all those 'other' questions now, it may very soon be too late.

The next steps we make in the study of human genetics are crucial. *Study* becomes the operative word. We must be very careful, because some of the possible applications and outcomes — new viruses, germline intervention, biological weapons, transgenic chimeras — may well be irreversible. And then our descendants will look back upon us as the last generation or two of genetically unaltered people who still had a choice, and a fair chance to know themselves as something more than the sum of their parts or their programming.

Here's an interim proposal: Let's not write the epitaph for *homo sapiens* before we have a little better idea just 'who' that most peculiar creature might really be.

The Side-Effects are the Main Effects —
The Great Unlearned Lesson of Media History
From Literacy to Genetic Engineering

W hy should I, ostensibly a scholar of culture and communications, concern myself with the *ethos* of genetic engineering? Isn't this somebody else's field?

I suppose it weighs on my conscience. I consider genetic engineering to be a rogue technology, far and away the most irresponsible application of a severely reductionistic information theory which — while it no longer has any credibility as a serious hermeneutic of language — still reigns supreme in the sciences. Because it is an indirect effect of the way digital technologies have restructured the theory and practice of communication, I regret that I did not see it coming. But that's no excuse.

Media history, like the history of ideas, amounts to second thoughts about things we take for granted. It is a surprising field of study because hindsight reveals so much that is hidden at the time new media technologies are introduced. When we first examine any medium of communication, we tend to see only its main effects. Phonetic writing seems to be a handy technique for inscribing the sounds of speech on clay, wax, papyrus, etc.; the printing press looks like an ingenious transformation of the wine press to reproduce unlimited copies of words and images; digital media appear merely to employ clever algorithms for combining ones and zeros to synthesise practically anything from music to words to the astonishing images from the Hubble Space telescope.

So what's missing? When we review the main physical or technical effects of any medium of communication, we tend to neglect the *context*, the psychological, social and cultural setting which each medium infects and influences in certain ways. We don't use our 'peripheral vision' — the sort of thing Marshall McLuhan was good at — to see the effects that a certain medium has on our senses, our sense of ourselves, and our culture at large. Indeed, it may be said that each new medium carries its own worldview, which can transform the self-understanding of the culture at large. "Culture is communication, and communication is culture," wrote Barrington Nevitt 20 years ago.[1] Because media of communication directly impinge upon the world of meaning, they can change the meaning of the entire world.

We may say then that besides its direct, immediate and obvious effects, every medium of communication has also its side-effects — often subtle, far-reaching (in time as well as space), and profound.

Speech

Whenever and however articulate speech first breaks the stillness, the wholeness of all that 'goes without saying' in human affairs until that moment, it also abandons the harmonies of music. The cacophony of human utterances can never again quite attain the 'ideal dialogue' of music, where many voices can 'speak' at once (or rather *sing*) without cancelling each other out. As George Steiner has said, "Our poetry is haunted by the music it has left behind." [2] The spoken word likewise leaves behind most of the body, except as accompanying gesture, and attenuates the body's connection with the Earth, indeed with time and space itself. The human voice alone, unaided by alphabet or archive, carries on beyond the speaker, even beyond the speaker's death. Human memory is no longer carried solely in the neurons of a single brain, but in all the remembered patterns of activity, customs, stories and sayings passed down the generations. The human mind, something more vast and pervasive than any single human being, is the offspring of language, the product of a shared cultural tradition. Until we speak — until we "share" in the word, as the *Rig Veda* (I, 164) first affirms — we each stand alone, secluded, burdened by minds that cannot unburden themselves. Nevertheless, speech itself opens up rifts and fissures in our experience: between I and thou, between then and now, here and there, singular and plural, etc., The spoken word joins together but also distances speaker and auditor, medium and subject matter. With the passing of years and the diligence of many speakers, this community of meaning can become a language, joining the many speakers into a single identity, yet at the same time rendering them unintelligible to speakers of another dialect.

Literacy

L iteracy is an abstraction from speech — 'literally' *pulling out* the bare sounds of speech from meaningful discourse to form phonetic letters, and 'figuratively' *pulling away* from the directness and immediacy of the original speech 'event' to the text. An organic whole — the union of speaker, spoken to (auditor), spoken about (subject matter) and spoken through (sound) — is broken up to yield a new medium, the written text, from which speech (or rather a simulacrum of actual speech) may be reproduced. This 'technique' is why Walter Ong discerns in phonetic literacy the foundational technology of the West; it embodies for the first time the basic strategy employed in every other technology: A meaningful 'whole' is broken up into meaningless 'units', which are then reassembled in preferred patterns. We need to bear this sequence in mind for all further developments which stem from phonetic literacy: first *analysis*, then *synthesis*... a new artifact produced by artifice.

The origins of writing are entirely pragmatic and commercial; the poetry and literary forms follow from simple accounting notations where first clay tokens were used by traders to stand for commodities, then pictograms to stand for the tokens, then the pictograms employed to represent the first syllable of the item — oxhead (*aleph/ alpha*), house (*beth/beta*), and so on. The Sumerians were the first to develop such a pictographic system, called cuneiform from their wedge-shaped stylus, a little before 3,000 BC. They also commenced from time to time using some of their thousands of pictograms (originally one for each word) phonetically. Egyptian culture evolved its hieroglyphs a little after the Sumerians, employing phonetic markers solely for foreign names. Elements of both systems survive

in the first workable alphabetic scripts, today known as Proto-
Canaanite, which emerged in Canaan and the Sinai about 1,700 BC
and consisted of 22 letters for the Semitic consonants only. These
early scripts led to the Phoenician alphabet, which broke away from
its Proto-Canaanite roots about 1,300 BC to become the ancestor not
only of Hebrew and Aramaic (which in turn begot the Persian,
Central Asian and Indian scripts), but of the early Arabic scripts as
well. When exactly the Greeks borrowed the alphabet from the well-
travelled Phoenician traders is a subject of some dispute; the transfer
may have occurred as early as 1,000 BC but is certainly visible on
inscriptions by 800 BC. The Greeks completed the alphabet by adding
letters for the sounds of the vowels. In fact, Greek is still the most
perfectly phonetic alphabet; modern English renders its 40 phonemes
(20 consonants and 20 vowel sounds) in only 26 letters, resulting in
many ambiguities, doubling up of vowels, etc.

   For our purposes now we need not go very far at all into the
detailed epigraphic history of literacy — well known in the works
of Naveh, Schmandt-Besserat, and others — to catch the principle
of *abstraction* by which phonetic literacy reassembles the world in
its own likeness. To make a long story very short, the entire world
begins to appear to the literate eye as (potentially) a perfectible text.
Rather than accepting the world as 'given', or indeed 'gift' as many
oral cultures do, literacy brings with it the challenge to 'edit' the
world, to remake it as one would craft a written text. The past need
no longer be repeated (lest it be forgotten), but can now be stored
on clay tablets or later in papyrus archives. The imaginable future,
to minds no longer compelled to recall every last detail of the past,
opens up as a vast new field of unexplored possibilities. Historical
consciousness, as R. Panikkar asserts, is born with literacy — not
so much as a record of the past, which can now be safely set aside,
but as this new thrust towards *the future*: the promised land, the
heaven above, the classless society, the good life for the children,
and so forth. Linear time — the time 'line' of the written text running
from a definite starting point to The End — drives out the old,
cyclical, recursive temporal rhythms of the pre-literate sensibility.
Now everything is a function of what it will *become* in the future,
'at the end of the line'. It is the incipient form of the evolutionary,

developmental worldview we so easily take for granted today.

McLuhan and Innis found in the tale of Cadmus sowing the dragon's teeth the founding myth of literacy. The old cyclical world of nature's rhythms, within whose coils humans must simply accept their fate, is transformed into a challenge that lies ahead, a conquest or discovery, a destiny to be fulfilled. The dragon is in the final analysis Nature herself; the 'teeth' taken from her slain corpse are the serrated, regular letters which spring to life in ranks and files like soldiers to do battle with that 'fate'. The origin of writing conjures up images of biting, dissecting, tearing and rending one world to render up a new one.

There are more ways than there are human mouths to produce the sound 'represented' by the letter 'a', but the single letter stands for them all. Literacy destroys the unity of the spoken word, yet literate culture seems ever thereafter impelled to seek some semblance of that lost unity. It can no longer be assumed, so it must be constructed: the Tower of Babel, the Universal Empire, the Global Market. Out of alphabetic literacy, as Robert Logan has emphasised, arise three further abstractions: monotheism, codified law, and the beginnings of science. There are many Gods — indeed, every mountain and river and tree houses its spirit for most traditional cultures — but the literate mind 'sees' that there must be one principle (of sacredness? of power?) underlying them all: YHWH. There are many human behaviours, customs, notions of right and wrong, rewards and punishments, but henceforward one written law will prevail: Hammurabi's Code, Mosaic Law, etc., There are many natural phenomena, but the mentality of literacy will seek 'laws of nature', and beneath them assume there to be a single principle — it is all water (Thales), atoms (Democritus), mind (Anaxagoras), indefinite (Anaximander), fire (Heraclitus), or, most abstract of all, Being (Parmenides). The very notion of cause-and-effect follows the sequence of a written text, from a beginning point through a differentiated sequence to a terminus. So begins the quest for the ultimate cause, the Prime Mover, the basic building block, the grand unifying formula... Today we call it science.

All this on the 'objective' side reminds us that with literacy comes a separation between humans and the natural world, observer and observed, 'subject' and 'object'. There is a novelty on the side of the 'subject' as well, symbolised by Socrates consulting his 'daimon': a new interiority, the dimension of self-reflection that comes with writing (as much a mode of meditation as it is a communications 'tool'); the "inner Man" of which Augustine spoke centuries later, awakens in this period. Karl Jaspers called it the 'axial period', which we today recall as much for the prominence of certain remarkable personalities as for its writings: Plato, Zoroaster, the Hebrew Prophets, the Buddha in India, Confucius and Lao-Tzu in China, and so on. The human personality emerges on a world historical stage already set and mentally furnished by literacy. The alphabet — 'a' is for the apple in Eden, too — is 'eaten', assimilated, embodied in the reflexive, self-conscious human subject. The innocence of oral/aural experience is broken, probably forever. Some scholars (McLuhan, Ong) make the most of the differences between oral and literate cultures so as to invoke a 'secondary' orality which might explain how today's electronic media have precipitated our departure from the literate — rational, sequential, conceptual — ideals of the Western tradition. Other writers sense a strong thread of continuity (we do still talk, do we not?), while still others (Illich and Sanders, for example) see the break as radical and insuperable (we can never go back). Most mainstream media historians (e.g., Postman) walk a line somewhere between the extremes.

All will, however, readily concede that literacy brings with it an avalanche of similar transformations which follow the same procedures of abstraction. Numeracy, for example: Each thing is unique, but we can count them all... Two apples plus two oranges do not add up to four of any- 'thing' until we ascend to a low plateau of abstraction which permits us to see them as four 'fruit'. And then of course there is money, invented about 650 BC by the alphabetic Greeks and Phoenicians. Every transaction between people is unique; barter and haggling over goods is a form of spoken dialogue. Literacy is the backdrop against which money emerges as the great equaliser, ultimately becoming an abstract monetary accounting system which will permit the exchange of practically anything for anything else.

In short, we recognise the mentality of literacy less in its banal direct effects — the impressions left by a stylus in wet clay — than by its dramatic side-effects. I would go so far as to declare that civilisation itself (literally 'city-culture', arising for the first time in Sumeria in tandem with literacy) is an unexpected concomitant of this new media technology. Writing is the first remote control device, so that taxes and tribute may be exacted from far provinces, so that records may be kept, so that larger and larger territories may be governed from one central city, so that armies of workers can be organised to assemble pyramids and city walls according to an architect's abstract designs (and further armies mobilised by enemies to knock them down). An elite of scribes and literate priests emerges, the literate 'knowledge class' replacing the shamans and elders of an earlier culture. A local, small-scale social order is entirely transformed, hierarchised into the familiar institutional forms of civilisation (temple, citadel, granary; i.e., church, state, treasury). Written, often secret, orders are passed on in one direction only, from the 'dictator' (the unlettered king literally 'dictates' policy to the scribes) down through this emerging bureaucracy to the rank and file of workers, soldiers, craftsmen and peasants, and so on. The chequered history of civilisation is all too familiar — "Civilisation originates in conquest abroad and repression at home," declared Stanley Diamond[3] — but at every step along the way, it is the sinews of literacy which flesh out the lineaments of empire.

The Manuscript Period

J ohn Ruskin considered himself a manual labourer because he wrote his books by hand. Handwriting (or chirography, following its Greek roots) is the main vehicle of literacy for well over two thousand years. From classical times right through to the Renaissance, only a minority of people could read and write — scribes and scholars, clerics and bureaucrats, children of the wealthy townspeople and the nobility. This long gestation period for literacy in the West had its strong points. Handwritten books were rare and expensive, so icons and images and the artistry that produced them were as important as words in conveying cultural traditions; music and architecture were communal art forms with far more cultural content than today's sterile skyscrapers and the 'muzak' in supermarkets. The medieval liberal arts curriculum in the universities nurtured not only textual literacy in the *trivium* (grammar, rhetoric, logic) but contextual 'pattern-recognition' strategies in the *quadrivium* (music, arithmetic, geometry, cosmology), without neglecting to pass on traditional crafts and practical skills in the mechanical arts. Today indeed, as the era of universal literacy recedes in most 'developed' countries from its peak in the 19th Century, we begin to appreciate the rich balance struck in the chirographic period between oral and literate forms of communication.

In a brief survey like this, we can scarcely pause to consider the persistence of Greek thought, of Roman administrative forms, of Christian beliefs, etc., in the transmission of culture from the ancient world to the modern. A single personality, exceptionally diligent but not especially brilliant or original, will have to stand here for the many strategies whereby books and the traditions of literacy survived the often violent vicissitudes of western history. Flavius Magnus Aurelius Cassiodorus many not be a familiar name to most students of media history, but in hindsight the contributions of this Roman 'Senator', as he styled himself, loom large indeed. His long life (c.490-580 AD) coincided with the tail-end of the Roman Empire. By dint of his father's position and his own organisational skills, along with some canny political footwork, he became *magister officiorum*, chief of the Roman civil service, just as this vast literate bureaucracy was brought under the heel of the unlettered Gothic chieftains who had taken over the Western Empire. As indispensable right-hand man to Theodoric and his successors in running the Empire, he not only saw the barbarians 'aping' Roman dress and customs, but foresaw a Dark Age fast approaching on the horizon.

Cassiodorus withdrew from public life in his fifties, and retired to his family estates in Squillace, well out of the political fray in the backward 'boot' of Southern Italy. There he founded two monasteries, Vivarium and Castellum, opening his doors to young scholars seeking refuge from the turmoil of the disintegrating Empire. Besides compiling a rather narrowly biased "Universal History" down to his own time, the "Instructions" he gave his monks included a simple directive which has echoed down the centuries. He told them, first, that their special duty was to acquire knowledge, both sacred and profane, and then enjoined them to begin copying and translating manuscripts from the Greek and Latin classics. He himself prepared a grammar of the 'proper' Latin he saw declining all around him, and was the first writer in the West to use the word *modernus* regularly. The task of the true 'modern', according to Cassiodorus, was "receiving and preserving some of the heritage of antiquity, ...discarding what is useless of the ancient, preserving into the new age useful treasures of the olden days." [4] He also made numerous compendia of classical works, while he and his monks avidly collected manuscripts wherever available. Their library at Vivarium eventually became a cornerstone of the Vatican manuscript archives.

Other monasteries took up this task of preserving and copying manuscripts; the nearby Benedictines may already have been at it in Cassiodorus' time. At any rate, all we know of the civilisation of the West before his time has come to us through those manuscript copies, enhanced by inventions of their own like the lower case letters you are reading now (Carolingian miniscules). Without the efforts of the monks over the long ensuing centuries of illiteracy, what we now call the High Middle Ages — the great cathedrals and schools of theology, Dante, Giotto, *et al.*, the first full flowering of indigenous European culture — would never even have occurred, and the Renaissance would have had no classical tradition — in the academy, literature, sculpture, architecture, etc. — to champion and revive. After the general destruction of Rome and the Roman world, as well as of the 400,000 scrolls in the Library at Alexandria in 640 AD, Irish monks and others in their scriptoria at the farflung corners of what was once the 'civilised' world kept up this quiet task, embellishing their copied books with all the phantasmagoria of their own archetypal imaginations. We owe those anonymous monks for whatever pretence we now make to be the 'civilised' inheritors of the Greek and Roman legacy. The 'Dark' Ages were not, as a flippant student once suggested, everything that happened before the invention of the light bulb, but simply a period when people forgot how to read and write and the achievements of the past receded from view. Literate monks from Ireland like Scotus Eriugena eventually re-educated the whole of Europe, as Thomas Cahill rightly maintains. In retrospect, the side-effects of Cassiodorus' instruction to make copies of the old manuscripts amount to very nearly the entire classical heritage of Western culture.

Print

The tendency of literacy to mould into linear sequences what the mind will accept as intelligible may well have obscured the obvious and observable fact that no medium of communication ever evolves in a strictly linear, or indeed predictable, manner. Lateral thinking is required, or else media history becomes a mere list of inventions, an account of their popular reception or rejection, plus perhaps a critique of their organisation and governance (or manipulation and abuse) by elites, opinion makers, capitalists, etc. The most telling example for a 'non-linear' approach to media history has always been the (exceedingly linear, sequential, segmented) medium of *print*. The crucial case turns of course not on Chinese and Korean handprinting techniques, or early block printing in Europe, but upon the invention by Gutenberg in about 1448 AD of moveable type and the mechanical printing press. The short form of this argument is that print intensifies all the effects of literacy sketched above, as well as adding some new twists in further dimensions of human affairs. Marshall McLuhan's pioneering analyses of print and the "Gutenberg Galaxy" of its side-effects in fact formed the founding subject-matter of media studies in the academy. These effects are so wide-ranging and pervasive, and so well-documented (in print yet), that taken all together they amount to nothing less that what we have come to know as the 'modern' world.

What would be left of the definitive innovations of the early modern period were we to delete the spin-offs from print? Probably chopping the ebb and flow of time (and the variably-sensed duration of human experiences) into standardised, unidirectional units by the mechanical clock, which Marx considered the first industrial tool; then the orthogonal organisation of space by the Renaissance contrivance of rectilinear perspective, mechanically representing three dimensions on a two-dimensional grid, which made the abstract subject-object dichotomy of the philosophers into an everyday sensory experience; and let us not forget gunpowder, always useful, plus a few improvements which pushed hearth and forge toward the eventual furnaces and engines of the industrial period. One might add Luca Pacioli's codification of double account book-keeping about 1500 AD, which began to organise commerce into the black and the red, profit and loss columns. But that's about it, and all of these inventions dovetail into the side-effects of print culture, so it's very difficult to segregate them.

The rudiments of later mechanisation are already present in the printing process itself. The intact original handwritten text is broken up into its individual letters which are then reassembled in regular rows on the galley for printing; the letters themselves are standard, interchangeable, re-useable parts; a division of labour is necessary (the text is manipulated separately by author, editor, compositor, printer, distributor, reader), as is the resulting specialisation of 'jobs' (a 'piece' of a work rather than a complete *opus* or *oeuvre*); indeed, the mechanics and demographics of mass production are put in place for all later industrial tools and products. The many mechanical processes which mimic the success of the printing press are well known, as is the consequent demise of craft guilds and hand production. These could still be considered among the 'main' or direct effects of the mechanical process of printing.

The novelty of the printed book itself (or perhaps first the penny paper) is quite simply staggering. It is in fact the first industrial commodity, the first 'product' in the marketplace of which every copy is identical to every other. Since this period when printing first transformed medieval Europe into 'modern' society, no traditional culture has ever coped with the intrusion of any such mass-produced 'commodities' without changing radically. A fixed price is for the first time possible, spelling the end of haggling and barter (vestiges of oral culture), and setting the stage for a recognisably 'modern' economy where identical 'units' or 'products' are exchanged for pre-set prices. Accompanying the transformation of the marketplace by this profusion of mass-produced commodities on the model of printing came a host of related social, political, scientific and religious changes. These are so well-known to historians of media that a brief summary may suffice here, relying on the extensive existing literature to justify the inevitable overstatements and simplifications:

• The 'author' is born, along with the notion of 'intellectual property'. The medieval 'author' might be a copyist, compiler, redactor or a commentator. But a printed book requires a consistent authorial 'voice', itself a somewhat artificial construct, and the economics of publishing that this author be considered the sole 'originator' of the text, and therefore its legal 'owner' (unless 'copyright' is ceded to the publisher). Copyright implies the attitude peculiar to the modern era that ideas may be treated as 'private property' — where exactly do ideas come from? — a print era notion increasingly difficult to perpetuate in today's 'digital' world.

• Vernacular languages (German, Italian, French, English, Spanish, etc., instead of Latin) come to the fore and gain dramatically in authority as local literatures and their oral epics are printed and made available to wide public audiences. Grammars and dictionaries are assembled, dissecting what were previously only oral tongues and bringing a new uniformity to 'preferred' usages and 'proper' spellings. New nationalisms of all sorts proliferate, and along with them a new 'democratic' spirit that will eventually transform the traditional oligarchies of Europe and require at least the appearance of the consent of the governed — informed at every turn by printed 'news' — in the political life of the newborn nations.

• This book becomes the first 'teaching machine', opening up public schools to teach ordinary people the skills of literacy and numeracy they will need in the new economy of mass production. Specialisation of disciplines is inevitable once reading matter proliferates; the medieval *studium generale* fades to a pious ideal in the face of this fragmentation of knowledge. And, as Ivan Illich has demonstrated, schools serve a further function as incubators for a mass consumer society. Beyond the particulars of book-learning, students absorb the 'hidden curriculum' that they are first of all 'consumers' of things produced by others. Auto-didacticism is discouraged, local knowledge and traditional lore discredited. The uniformity and homogeneity of the printed page underpins the mental and social conformity of the emerging bourgeois epoch. A mass society fosters the illusion of 'individualism', all the while cultivating standardised responses to standard stimuli (textbooks, adverts, propaganda, etc.).

• Religious reform in Europe begins with a perceived abuse of printing. The Church had found a profitable way to employ the printing press: They printed tickets to Heaven, called 'indulgences', with which a succession of Popes hoped to pay for their incessant wars, and Julius II for his ambitious project to rebuild Rome to its former glory. Gutenberg's own first press in Mainz cranked out lucrative indulgences by the thousands. In the next century, Martin Luther denounced the practice, indeed the entire affluent ecclesiastical hierarchy, and outflanked the Church's old-fashioned oral methods of preaching with a spate of printed pamphlets that circulated quickly to wide acceptance (his books were the world's first bestsellers). Luther and his fellow 'Protest-ants' began translating and printing the Bible in vernacular languages which anybody could read and understand for themselves without the intervention of the Roman Church or priesthood. Decades of war ensued, but the Reformation redrew the religious and political map of Europe and remains a powerful legacy of the print era.

Print culture fostered not only the scientific attitude, but scientific method itself. The dissemination of ideas in print certainly encouraged people to take a fresh look at the world around them, and emboldened them to take issue with traditional authorities whenever these flew in the face of direct observation. Galileo's famous cry, "I have seen it with my own eyes!," exemplifies this confident new inquisitive spirit. With print also came the alphabetic index for classifying all the new data, a novel method of organising information which eventuated in the encyclopedia. Moreover, as Elizabeth Eisenstein has extensively documented, scientific publications proliferate during this period, so that new evidence from many sources can be compared, reviewed and evaluated. Crucial to this new openness to the acquisition of knowledge is a new technique for verifying observation, one tied directly to the process of printing a 'proof' sheet to check for and correct typographical errors. For oral cultures, once upon a time, personal *experience* was the touchstone of truth. For literate culture from the classical through to the medieval periods, the locus of truth shifted from the experience to its written *expression*, which must obey abstract laws of logic and survive the rigors of dialectical argument to be considered true. For the emerging scientific mentality, however, what counts is the *experiment*, which becomes the cornerstone of true and reliable evidence in the new edifice of 'objective' knowledge. What is an experiment? A result obtained by procedures which may be — must be — repeated to be verified. The criterion of truth by experiment is its *repeatability* (by others who have read the published results, etc.). It would overstate the case to say that science replaces faith in God by the *certainty* it finds in results obtained by experimental method, but not by much.

The Control Paradigm

In both the galley 'proof' and the proofs by experimental method, we catch sight of something which will eventually determine the entire tenor of the modern age, and may serve here to summarise in a single word the multifarious side-effects of print culture. It is the mechanism of *control*, a technique rapidly applied to every aspect of social life and, following Francis Bacon's admonition to "conquer and subdue" Nature, to the exploitation of the natural world as well. It is to empirical knowledge derived from experiment that Bacon refers when he declares: "Knowledge is power."

The noun 'control' turns into a verb, meaning 'to oversee', just as it comes over into English in about 1600 AD (much as the noun 'access' metamorphosed into a verb about 1990, in response to the Internet). Until then, 'a control' was in French a *contre-roll* (or earlier in Latin *contra-roulada*), a 'counter-roll' or duplicate register on vellum or papyrus. This old fashioned kind of 'control' was simply a duplicate list, or a master list — of taxpayers, soldiers, parishioners, students, etc. — which both preserved and, if need be, verified the original 'roll'. The 'control' was kept at Rome, or at army headquarters, the treasury, registry, etc., as any similar bureaucracy keeps a 'file copy' today, sometimes to the chagrin of local officials trying to fiddle the accounts. Our old friend Cassiodorus would have pored over hundreds of such 'rolls' in the normal course of his duties as chief of the Roman civil service.

In the age of print, the galley of type (or the metal 'chase' of an entire page) serves as the ultimate 'control'. Type is composed as a mechanical 'copy' of a handwritten text, but once the typeset text exists it utterly supplants the original.[5] If you own the galleys, you 'control' all possible printings, much as in a later era the photographic negative becomes the 'control' for any positives printed from it. This dynamic is simple enough, but the latent tension between copy and original soon becomes open conflict, then a fight to the death. In this age of burgeoning mass production, the copy takes over; the mass-produced replica begins to supplant the original in importance and authority. Printing is the template for practically every ensuing industrial mass production process. And there is money to be made here. Mass production means that a single woodcut by Dürer, once printed, can be sold ten thousand times, each time for a fraction of its price as a 'one-off', and yet eventually earning more than the single picture ever would. Mass production has its appeals. Now everybody can afford a printed woodcut by the great artist, not just the rich. Depending on your perspective you will argue that this either raises the general level of culture, or debases Dürer to the popular level. At any rate, in the print era things become more valuable as commodities the more easily and rapidly they can be replicated by mechanical processes. The downside is that, soon enough, *only* things which can be mass-produced are fabricated at all. Traditional arts and crafts swiftly decline in favour of industrial processes. Before long only things mechanically 'controlled' in this manner exist in the marketplace; the story is a familiar one.

But the effect runs riot well beyond the economic sphere; indeed, the urge to bring everything in the world under 'control' seems unstoppable. All of Nature comes to be perceived as a machine, which can and should be brought under human control. It is the world Francis Bacon proposed in the early 1600s: politically, under the control of an absolutist empire; scientifically, under the control of inductive experimental method and technocratic 'development'; psychologically, under the control of the 'rational' ego and its not-so-rational will-to-power. Today, this urge to control seems to have gotten entirely out of control, or maybe encountered its natural limits, as it attempts to dominate matter itself in the nucleus of the atom and the innermost recesses of life in the DNA of the living cell. That this 'urge to control', for all its well-known pathological excesses, still has us in its grip today might be considered the most surprising side-effect of five centuries of print culture.

Yet there was, even early in the print era, always a deep-seated uneasiness about it all. Is this kind of technology a devil's bargain? Are the newfound powers it brings to manipulate both the social and the material world a blessing, or a curse? A profound disquiet is expressed in the literature of the period, just as it is in today's science fiction. Theologically, the problem of copy versus original appears in the dynamic between creature and creator. The Judaeo-Christian prototype is Lucifer, the angel 'closest' to God, whose sin is *autarchy*, the claim of a created being to be his own *arché*, his own source, his own creator. The intersecting theme in the Hellenic tradition first appears in the founding myth of technology, the story of the titan Prometheus, the tragic hero 'most like' Zeus himself in Hesiod's *Theogony*, who steals the divine fire from the forge of Hephaistos — the very 'spark' of life — and is punished by the Gods for his sin of *presumption*. Philosophically, the question becomes whether the ultimate structure of things is heteronomic (ruled by an other) or autonomic (self-ruled), and spins off into the similar problematics of the One versus the Many, the whole versus the part, and the master/slave dialectic analysed by Hegel. In short: Either you control 'it', or it controls you. Everywhere during the print era, old heteronomic orders (church, state, society, property, tradition, etc.) are challenged by the new spirit of autonomy. The copy, the control mechanism, is in the ascendancy. As the Many (copies, creatures, citizens, etc.) proliferate, they challenge and eventually overwhelm the One (original, creator, king, etc.).

It is in literature that the most disturbing and persistent echoes may be heard, issuing from subterranean currents both social and psychological which run deep below the surface confidence of the era and spill over right into our own troubled time. First there is the story of Faust, who sells his soul (in this era just another commodity) to the devil for knowledge and power. It is often overlooked that the ironic character of Faust's rebellion against God is strongest not in the story's literary reworkings by Marlowe or Goethe, but in the early puppet plays for children, popular for three hundred years, where you can actually see the strings from above animating the actions of the little marionette. Faust is originally the puppet who wants to cut the strings, the tension between creature and creator visible to all. The motif of their stormy relationship reappears arrayed in all the rhetorical thunder and vivid lightning flashes of Milton's *Paradise Lost*, Blake's *Book of Job* engravings, or Shelley's *Prometheus Unbound*, and yet attains what turns out to be its definitive modern expression in the chilling vision of a talented young woman. Mary Shelley's *Frankenstein, or The Modern Prometheus*, is not only the most successful novel ever written by a teenager, but the first real science fiction story. It comes about because Mary manages to fuse two primal Gothic horror stories: the story of the creator who recoils in horror from his monstrous creation, depicted as a second Adam who ends by destroying Frankenstein and all that he loves; and the even more poignant story of the creature abandoned by its creator, totally alone in a hostile alien world. The morality tale of Frankenstein, perhaps the single most re-made cinematic motif of the 20th Century, nowadays simply stands for the would-be controller whose creation escapes his control. It is the nightmare that persists beneath the 'rational' dream of unbridled technical progress. In Goethe's *Faust, Part II*, composed in the same decade as Mary Shelley's *Frankenstein*, there is a 'sorcerer's apprentice' episode where Faust's assistant Wagner creates an artificial human, Homunculus, in the laboratory. The sardonic Mephistopheles' final words on Homunculus highlight what may be the most unsettling side-effect of the very powerful technologies unleashed during the print era: "In the end we are dependent on the creatures we ourselves have made'.[6]

Electricity

			.-- W
.-- W	.... H	--. G	.-. R
.... H	.- A	--- O	--- O
.- A	- T	-.. D	..- U
- T	.... H		--. G
			.... H
			- T

Is it a question, or an archaic exclamation? No punctuation was recorded at the time; the 18 letters seem to have been chosen mainly to test various combinations in the system of alternating dots and dashes. Either way, there is little question what Samuel F.B. Morse had wrought by sending this first 'long distance' telegraphic transmission between Washington DC and Baltimore (37 miles) on May 24, 1844. His invention inaugurated a communications revolution, the transmission of language (and later images) spelled out in electrical impulses. Morse's achievement was soon duplicated by Cooke and Wheatstone in England, using a cumbersome system that Morse's apparatus and code had already obsolesced. Morse and his partners devoted long years to refining the technical details, cajoling the government for money, and working out a code which did not require a codebook to decipher. "Morse Code" was based directly upon the 26 letters of the alphabet; indeed it was the number of letters in a printer's box of type that told them which were used most often, so that in Morse's system these common letters require the least dots and dashes — 'e' a single dot, 't' a single dash, etc.

The idea of transmitting messages by electrically transposing the already abstract letters of the alphabet was first publicly broached in a mysterious letter entitled "An Expeditious Method of Conveying Intelligence," published February 17, 1753 in the *Scots' Magazine* by a certain "C.M." (never identified), who suggested 26 wires —

one for each letter of the alphabet. In the 90 years between that first
notion and Morse's success, at least 60 abortive attempts were made
to construct a device to send messages electrically. Tom Standage
has detailed some of the more interesting ones, as well as the
subsequent history of Morse's invention, with journalistic flair. By
the time the first transatlantic cable was completed in August of
1858, praise for the new 'connected' world reverberated from the
popular press to the journals of Emerson:

Thought's new-found path
Shall supplement henceforth all trodden ways,
Match God's equator with a zone of art,
And lift man's public action to a height
Worthy the enormous cloud of witness,
When linked hemispheres attest his deed. ...

This feast of wit, this triumph of mankind;...
Urging astonished Chaos with a thrill
To be a brain, or serve the brain of man...
Mind wakes a new-born giant from her sleep.
Twirl the old wheels! Time takes fresh start again,
On for a thousand years of genius more.[7]

Indeed, this first flush of enthusiasm mirrors to some extent all
the early enthusiasm for the Internet a decade ago: "It is impossible
that old prejudices and hostilities should longer exist, while such an
instrument has been created for the exchange of thought between
all the nations of the earth."[8] But the undersea cable failed in October
for lack of adequate insulation, and only got properly underway in
1866. And the deployment of the telegraph, again much like the
Internet, did not issue in a new era of world peace, but followed a
more familiar and mundane path: "At its very birth, the telegraph
system became the handmaiden of commerce."[9]

"Thought's new-found path" through the electrical current was
poorly understood until James Clerk Maxwell first tentatively
combined equations for magnetism with those for electricity to
formulate electromagnetic theory in 1864, later consolidated in his

Electricity and Magnetism (1873). The combination of a working device and a workable theory opened up an astonishing range of practical applications, most of which have turned into today's commonplace electronic media of communication: Edison soon produced both the phonograph (1877) and the incandescent lamp (1878), Bell followed with the automatic telephone (1887), then Marconi achieved first simple telegraphy without wires (1895) and eventually the first transatlantic radio telegraph message (1901). By 1902, radio telegrams were a quotidian commercial reality, photos were being sent by wire as well as words, and the following year the Pacific cable permitted a message to be sent around the world in 12 minutes. Our main concern here is with communications media, but we should mention in passing that Einstein's Special Theory of Relativity (in which he asks what happens if you are travelling at the velocity of light) was published in 1906, while Millikan confirmed the existence of the electron, the basic 'unit' of electricity, by 1908. Light, the visible portion of the electromagnetic spectrum, which had already been used by Daguerre to make photographs (1837), was now almost magically transformed by Edison into moving pictures, advancing in scarcely 20 years from the kinetograph, the first true motion picture camera (1889) to kinemacolour movies by 1911. The first transcontinental phone service in the USA was up and running by 1915, as well as the high vacuum radio tube (Langmuir) and tuned radio frequency reception (Alexanderson) which together would help produce the first commercial radio broadcast of voice in 1920. Radar was on the scene by 1922, talking movies by 1926, and Crooks' cathode ray (discovered 1878) had been put to use independently by Zworkin and Philo Farnsworth to develop the first working television (in the lab only) by 1927.

We may pause to catch our breath. This quick succession of inventions should alert us to an important novelty, namely the dizzying *acceleration* of developments in the media of communication over the past century. It is literally mind-boggling, and it has by no means slowed down since Bell Laboratories invented the transistor in 1947, and Noyce and Kilby each produced their own versions of the microchip in 1959. While telegraphy and telephony can be seen as a direct application of the analytic technique of literacy — breaking something (the human voice, for example) down into bits and then reassembling it into preferred patterns at the other end — the net effect of all these innovations is a world where communication moves at the speed of light, filling the air with patterns undetectable by normal human senses. It is not overstating the case to say that over the course of about six decades, one might observe an entire cosmology collapsing and a new cosmology in the throes of birth. In the early years of the 20th Century, the intellectual atmosphere was 'charged' with the 'electrifying' news that the entire universe of our experience consists of *energy*, either concentrated as matter or dispersed as radiation. Instead of Newton's universe, where the norm was a body 'at rest', the norm in Einstein's universe is the speed of light expanding in a spherical wave at 186,000 miles per second. This is a universe of flux, transformation, metamorphosis; the mechanistic 'essence' of the print era gives way to the ever-changing 'existence' of the electronic world; the universe no longer looks like a stable gravitational 'system', but like a constantly unfolding 'scenario'. This new cosmology is a world of 'events' in constant interaction, not a static world of 'things'. Things just won't sit still any longer: instead, they 'happen'.

"The artist," Marshall McLuhan famously observed, "picks up the message of cultural and technological challenge decades before its transforming impact occurs. He... builds models or Noah's arks for facing the change at hand." [10] This is emphatically what 'modernism' meant to many of the poets and artists who sallied forth under that banner in the early years of the 20th Century. It meant taking leave not only of Victorian fustian and representational imagery, but taking leave as well of the static, mechanistic Newtonian universe and immersing oneself in the organic rhythms of the dynamic, wholly energetic universe heralded by Einstein's famous little equation. One might cite jazz and modern dance as examples, or even the no-longer-static arts of sculpture and architecture, but the fundamental assumptions of literacy are in play here as well. Ezra Pound, the great pioneer of 'open' verse, spoke of artists as "the antennae of the race," tuning into "a world of moving energies."

"Energy creates pattern," he declared: "We might come to believe that the thing that matters in art is a sort of energy, something more or less like electricity or radioactivity, a force transfusing, welding and unifying. A force rather like water when it spurts up through very bright sand and sets it in swift motion. You may make what image you like." [11] As if to present just such images, the visual artists — Impressionist, Dadaist, Cubist, Vorticist, Surrealist, each in their own way — tore up the rule-book of representational imagery from the Renaissance and began exploring new dimensions of a liberated visual imagination where 'primitive' art could meet 'civilised' images face-to-face, where everyday objects became *objets d'art,* where Western canons of taste and proportion were deliberately violated, and the Mona Lisa seemed amused to sport Duchamps' naughty moustache. (One is sometimes tempted to think that today's 'post-modern' theories appeal most to those who never really appreciated what the 'modernists' were up to half a century earlier). Norman O. Brown used to remark that novelist James Joyce fissioned the etyms, the roots of words, some years before Enrico Fermi fissioned the atom, the supposedly solid 'substance' underlying all western substance metaphysics for 2,500 years — which had evanesced in the new physics first into a miniature solar system (Bohr, 1913), then into a little galaxy of its own. In short, the new electric communications technologies bore with them a new cosmology, indeed the most thoroughgoing shift in worldview since the print era scientists cast aside the medieval, theocentric universe and set about designing one of their own on purely rational, empirical, and quantifiable terms.

It was a new world, or at least it should have been. But the birth of an organic worldview heralded by so many thinkers and artists of the 20th Century (and easily tied to the rise of the 'counterculture' ecological sensibility in the 1960s) was aborted, or maybe the 'connection' somehow shortcircuited between those visionaries and the mainstream society. After all, it's 500 years since Copernicus and most of us still 'see' the sun rising and setting, even though we 'know' it's the Earth that's moving. What has happened in this case (often under intense pressures, World Wars, etc.) is that the new communication technologies were hastily so to speak 'pasted' or 'tacked' loosely onto the structures and institutions of the print era. A wholly unforeseen side-effect of electronic communications media has been not only their basic incompatibility with the older cosmology, but the persistence and inertia of the stolid formal structures stemming from print culture: e.g., individualism, intellectual 'property', privacy, etc. The Einsteinian worldview was blithely declared to pertain only to the very big or the very small — galaxies or atoms — while good old Newton would continue to suit us for everyday business-as-usual here in Everytown.

But the fit between the two has never been good, and seems to be getting rapidly unworkable. Electronic media have a 'genius' (or would you call it a 'daemon'?) of their own, radically at odds with the sequential, one-step-at-a-time, segmented character of print culture. Lately something about this jerry-built arrangement seems to have come 'unstuck'. Something is happening 'all-at-once' to most of our most trusted and venerable social structures, though it's tricky to assess or predict the effects of new technologies when you're living smack in the midst of them. But the 'hybrid' arrangement which employed electronic media to revamp print era institutions has led instead to a peculiar kind of myopic giantism. Certain institutions have become vastly more powerful and pervasive, but at the same time begun to fatally threaten the very values they were instituted to preserve. Certainly the 500 year-old 'modern' institutions of print culture — e.g., the nation-state, the economy, the school system, the limited liability corporation, the medical establishment, the research culture of science, etc. — have all been transformed by these new technologies, and often all to the good. Such innovations have surely helped to make a profit, or defend a nation, or cure an illness, or at least contributed to learning. But in some instances, the effect of grafting the relativistic Einsteinian worldview onto print era institutions has unstrung the latter entirely and threatens to be catastrophic for people and the planet alike. We should not hesitate to point out institutions which have today reached such a crisis point, even if it means we must run a little ahead of our chronological exposition (a comfortable print era sequence) to cite only the most blatant instances:

• Once armed with nuclear weapons (1945 *et seq.*), the nation-state becomes a malignant atavism. Under international law, the nation-state has the right to defend itself by the use of any and all weapons at its disposal. In the nuclear age of 'mutually assured destruction', this is a recipe for disaster. The means a nation-state will have to use to preserve itself will ultimately destroy it, or at least most of its people. As de Kerckhove notes, the bomb is useless as a tool or a weapon. Its only function is symbolic: "The greatest communication medium mankind has ever invented," says he, "not for information but for transformation." [12] Yet we enter a new century with ever more nuclear-armed nations; just how long will it take us to get the message?

• Economy used to mean household management, then regional self-sufficiency. No more. Gigantic corporate pyramids and multinational behemoths bestride the lanes of commerce in today's 'wired' world, yet seem blind to the fact that their machinations have plundered the resources and degraded the environment which underpins their very existence. Today's 'global marketplace' (existing mainly in cyberspace; where five out of every six dollars are traded daily) has already destroyed the diverse regional economies of the wealthy countries, and is now attempting to colonise the last remnants of the vernacular sphere in the so-called 'Third World' under the guise of 'development' and 'cultural production'.

• Superimposing the new cosmology onto the control paradigm in science has culminated in the discovery and deployment of practical techniques for genetic engineering, which *in toto* amount to nothing less than the attempt to control all life by rational programming. 'Patent it, package it and peddle it' runs the rubric for today's 'knowledge economy'. Genetically engineered organisms are made-to-order 'commodities' for this new gold rush, even though they may very well portend the end of 'Nature' altogether. The obsolete nation-state is living on borrowed time as a trade and immigration barrier to the 'borderless' world of free trade, yet seems quite powerless to prevent the entire planet from becoming one gigantic biotech lab.

So how did we come — and come so quickly — to such a fateful impasse? We have the means to communicate with anybody on the planet in seconds, yet we readily demonise other cultures and begin to fear even our neighbours. The mechanisms of control we have put in place to help us communicate — and thereby to feed, clothe, house and protect ourselves — have taken on a life of their own and brought us all under their sway. We see more deeply into the atomic structure of matter or the intimate genetic processes of life itself than any generation before us, and yet we are aware that we may not be able to keep the new powers bestowed by this knowledge from doing away with us entirely. It is the *automatism* of our over-organised society that now threatens to extinguish not only one institution or another, one freedom after another, but the continued viability of human life on Earth. As Vaclev Havel recently observed, "Today the most important thing... is to study the reasons why humankind does nothing to avert the threats about which it knows so much, and why it allows itself to be carried onward by some kind of perpetual motion." [13]

The Automation Idea

The idea of automation is not new, although its realisation is mainly a 20th Century phenomenon. There are hints of the notion as early as Homer's *Iliad*, and it appears full-blown in the ancient legend of Talos, the giant automaton of brass said to have once patrolled the shores of Crete. A Buddhist legend at least a thousand years old — and probably much older, since it is set in the time of the Emperor Ashoka (Third Century BC) — relates that in "the Kingdom of Roma" there were "spirit-bearing engines" (*bhuta-vahana-yantra*) which move under their own power, as well as mechanical men (*yanta-purisha*) used as instruments of defense. R. Panikkar comments pointedly:

"These robots are autonomous; they follow their own rules. They may have been planned or constructed by human beings, but they are not subject to them and are in a way superior. Nobody can overpower them. Their engineers can only steer them, and this only under certain conditions. Once let loose, even their constructors cannot control them. Once you split the atom... " [14]

So from the earliest days, other cultures perceive automation coming from the West, from the great and powerful "Kingdom of Roma." In the Middle Ages, automation stepped out of legend and into daily life, at least at the conceptual level. The Church had been under fire for dissolute clerics, and people wondered aloud whether the Mass would still have any salvific value if celebrated by a drunk or a lecher. The response of the Church hierarchy was to declare the ritual of the Mass *ex opere operantis*, that is, 'an operation that operates independently of its operator'. Such issues were subjects of vigorous debate at the time. In the same period, for instance, the Second Lateran Council (1139) declared the crossbow illegal in warfare against Christians, because human life should not be taken 'automatically' by a mechanical device, namely the new invention of the trigger. It was, however, deemed acceptable to shoot Arabs with crossbows.

Leaving aside this fascinating, and distinctively Western, history of the idea, the first fully automated device — a self-regulating machine functioning independently of its operator — had to wait until the 19th Century invention of the thermostat. The rising/falling mercury in the thermometer makes or breaks a circuit for an electrical switch which turns the furnace on and off at a pre-set temperature. The one device 'controls' the other. Simple enough, but as ever more such devices are perfected the ramifications embroil all of us in complex consequences with which we still struggle today.

The relevant inventions include the punch cards that were used to regulate the weaving patterns of the automated Jacquard loom (perfected by 1804), similar in principle to the revolving drum in a player-piano. Charles Babbage actually designed a functional 'calculating engine' (a successor to those of Pascal and Leibnitz) in the 1830s, but it was never built. The earliest working forerunner to the computer was a tabulator Herman Hollerith put together at the turn of the 20th Century to collate the first US census. It was really just a glorified adding machine, but combined with a statistical component and a logical sequencer it was a giant step toward the age of artificial intelligence. Hollerith's little company merged with the ones that produced those other components, and he became the first president of IBM. Soon enough the company was wholly identified with its next CEO, Thomas Watson, who had no qualms about offering the technology to the Nazis for the ethnic profiling requested in the census they ordered immediately upon assuming power in the early 1930s. Edwin Black has documented the subsequent history of IBM's deep involvement with the Nazis in his shocking book *IBM and the Holocaust*. IBM personnel and their machines (never sold; always leased, as they are today, always programmed and serviced by "the solutions company") went right into the death camps; those are Hollerith punch card numbers on the forearms of the Holocaust victims, classifying them by age, nationality, ethnicity, able-bodiedness and so on. We should not get sidetracked by this atrocious first application of automation to the mechanisms of mass murder, but we should not leave the dark beginnings of the computer age unremarked either. In any case, the idea began to take hold, ready for mathematician John von Neumann to construct the first true computers, the room-sized vacuum tube calculators he built for the Manhattan Project in the late 1940s — first the ENIAC, then the all too appropriately acronymed MANIAC — designed for the sole purpose of completing the calculations for Edward Teller's effort to build the first H-bomb. Without von Neumann's 'calculating engines', it has been estimated that there would not have been enough labour-intensive hours left in the 20th Century to do by hand the complex computations needed for what C.G. Jung once called "that peculiar flower of human ingenuity, the hydrogen bomb." [15] No one then envisaged personal computers, or the 'information society', but one man had more than a clue where it all was going; he had a method...

In MIT mathematician Norbert Wiener's *The Human Use of Human Beings — Cybernetics and Society* (1950), the automation idea at last comes of age. Here Wiener presents in summary form the approach to organisation emerging in the electronic age, a general theory called *cybernetics* (from the Greek *kybernetes*, a 'steersman') which is intended to explain how phenomena maintain themselves through information and feedback. Wiener explicitly applies the theory to both machines and living organisms, highlighting the way systems may provide information to themselves and thereby mimic purposeful behaviour. (We shall return to this theme, as it inaugurates the use of engineering models to describe biological processes, the very basis of genetic 'engineering'). This circularity — which once and for all displaces the linear, cause-and-effect mechanisms of the print era — operates through "negative feedback," by which a system is able to sample its environment (as the thermometer samples the surrounding air for the thermostat) and correct itself automatically, to meet an expected and predictable result. Jeremy Rifkin extrapolates:

"For Wiener, all purposeful behaviour reduces to "information processing." He wrote, 'It becomes plausible that information... belongs among the great concepts of science such as matter, energy and electric charge. Our adjustment to the world around us depends upon the informational windows that our senses provide.' ...Wiener came to view cybernetics as both a unifying theory and a methodological tool for reorganising the entire world. Succeeding generations of scientists and engineers concurred. With the aid of the computer, cybernetics has become the primary methodological approach for organising economic and social activity. Virtually every activity of importance in today's society is being brought under the control of cybernetic principles." [16]

Wiener's cybernetics is mathematically based upon the 'transportation' theory of information transmission produced by Claude Shannon (of Bell Laboratories) and Warren Weaver in their book from the previous year, *The Mathematical Theory of Communication* (1949), a simplistic model which attempts to match input transmitted through a telecommunications channel to output received at a destination with a minimum of 'noise'. From a humanistic point of view, it is a bizarre theory in many respects — in particular because it has nothing to do with *meaning*, which is in fact held to *reduce* the information potential of the system — but it remains today the basis of information theory, and the most widely accepted model of communication for the sciences in general. Following Barrington Nevitt's analysis, "Shannon created a paradigm for machine communication. Warren Weaver extended it logically to... any communication whatsoever." [17]

Because it is so blindingly obvious, the sender/receiver model is both easy to critique and difficult to refute. Yes, in any act of communication, there is a sender and there is a receiver. This is a little like saying that any geometrical solid has both height and width. Of course, but the resulting two-dimensional picture is not only flat but flawed. The real figure also has depth (volume), it exists in *time*, and it has both intrinsic and extrinsic relations with other geometrical figures. Similarly, there is in any human communication an interiority, the dimension of meaning, untouched by the sender/receiver model. There is always also the medium, which as we have seen shapes and inflects the character of whatever may be communicated. Time enters any human picture of communication too, and in a profound way, since sender and receiver alike are *changed* in and by the very act of communicating with one another. Moreover, there are undeniable intrinsic (pre-existing) and extrinsic (actual or possible) relationships between human beings, which a machine model of communication will either simplify to an absurd degree or ignore entirely.

In tandem, the Shannon/Weaver model and Norbert Wiener's cybernetics — both notions now over half a century old — have brought us to construct around ourselves a society in the grips of integrated, automated 'systems' of control which regulate not only their own behaviours, but most human activities as well. To see the automated technocratic state in full bloom, you need only visit a modern hospital, international airport, or 'defence' facility. In the sixties, McLuhan and his followers spoke of communications media as "extensions of Man," extrapolations of the physical human body up to and including the ethereal projection of our nervous systems into the ever more complex 'neutral nets' now girdling the globe — radio, television, computer networks, satellite telecommunications, etc. What image of the human being does this "extension of Man" into electronic media conjure up today? As recently as the Renaissance, the human being was viewed as a 'microcosm', a little universe which faithfully reflected every aspect and dimension of the living cosmos at large. Today we must instead view our information society as a 'Macanthropos' — a gigantic, autonomous human facsimile that has taken on a seemingly purposive life of its own, well beyond the control of its would-be controllers. Little did anyone suspect that the automation idea would end by automating the entire culture that had conceived it. Mary Shelley worried about science making a monster it couldn't control. Our Frankenstein monster has already arrived, in the shape of a technologically mediated environment that surrounds and envelops us, and within which we are increasingly hard-pressed to figure out how to survive as human beings.

The Digital Revolution

The history of the subsequent digital revolution could be described, at every important step along the way, as a series of failures to detect or predict the astonishing side-effects of digitising words and images. As we have seen, the first computers were built for state or military purposes. IBM (International *Business* Machines) extended the computer into business as a tool for tighter organisational control, but failed to anticipate demand for the personal computer. The manufacturers tied to corporate concerns also failed to anticipate the 'mouse' and the ease of use which this first consumer application of elementary robotics offered in manipulating things on 'the other side' of the screen. By and large, the early computer manufacturers simply did not see past the hardware they were selling, allowing software companies like Microsoft to capture and retain the cutting edge of the industry. The decentralised DARPA net was set up in 1969 to guarantee military communications after a nuclear attack (forgetting that EMP would burn out all the chips anyway); but no one then anticipated that this 'net' would soon combine with others to 'morph' into the World Wide Web and the other components (e-mail, chatrooms, etc.) which go to make up the Internet, today boasting half a billion users. And all the very literate, visually-orientated commentators on the Internet missed what quite suddenly appears to be the obvious next step — the emergence of mobile phones (Nokia), and the swift convergence of technologies which permit Net access through these phones (Samsung), possibly obsolescing the clunky 19th Century typewriter keyboard in the process. Given the rapid refinements of voice recognition programmes, Myerson of Cambridge maintains that the mobile phone may very soon swallow the Net entirely.[18] In Asian countries, not hemmed in by the strictures of the phonetic alphabet, it seems to be happening already. The entire development of the information society has been one such unexpected, unpredictable side-effect after another.

Binary code, anticipated by Leibnitz as far back as the 17th Century, came into its own with electronic communications, particularly the inventions of the transistor (1947) and the integrated circuit (1959), which made it practical to apply on a wide scale. Binary is one crucial step simpler than Morse's trinary code (dot, dash *and* space); it breaks down any electrical signal for transmitting information into its simplest on/off components, which register in a coded programme as a long string of ones and zeros. I am not competent to get into the debate as to which algorithms proved decisive in working out and applying the spectrum of mathematical laws which today govern the systems designed to communicate or manipulate information. I must also leave at a mere mention the famous effort of British cryptographers during World War Two to decipher the ENIGMA code used by the Nazis, which brought the names of Türing and others to prominence. (Oddly enough, it has recently come to light that the Nazis used the ENIGMA machine right off the shelf, with the code on file in the British Patent Office all the while. The great 'decoding project' went doggedly ahead in the belief that those Nazis must be cleverer than that...) These things have an uncanny momentum to them, which I have elsewhere referred to as *the technological imperative*: If it can be done, it must be done (before the other guy does it first). If not one inventor or theoretician, then another soon comes up with the relevant idea or programme or innovation. The very process of invention seems to have an automatism to it, or at least a certain inevitability. The cybernetic principles of information theory brought into existence the basic hardware, but it is through the coded programmes of its software that the computer has been turned into a kind of 'universal machine' with the flexibility to automate not only the reproduction of words and images, but practically every aspect of human life in what Arnold Toynbee, of all people, was the first to label "The Post-Modern Age."

No aspect of contemporary life has been untouched by these changes — from the global economy and the rise of transnational corporations, to the state bureaucracy and defence establishment, to schooling and the home environment, mass media, healthcare, entertainment, what have you. Regarding every single one of these transformations there is already a voluminous literature, well beyond our scope to reduplicate here. Perhaps the most significant overall shift has been in the very nature of the capitalist enterprise itself, from an age of ownership and private property and commodities to what Rifkin is now calling "The Age of Access." First marketers of 'commodities' were transformed into providers of lifelong 'services'; today they are fast mutating into marketers of a vast range of programmed, simulated life 'experiences' — not just computer games and 'virtual' reality, but theme parks, 'reality' TV, travel 'packages', and so forth. You can now visit a sanitised 'Venice' in Las Vegas, and sample 'the California experience' in a Disneyland theme park, complete with redwood trees. Everything in the world can be turned into a commodity for sale; 'cultural production' is today's codeword for ripping off bits and pieces of the arts and culture from their traditional community settings so as to package and peddle them as 'entertainment' in cyberspace. The music videos of MTV set the pace for cultural production pitched into the gaping maw of the global marketplace, where there is no longer any discernible difference between programme, product and advertisement. The critical distinction between *creativity* (which relies on spontaneity; its model divine inspiration) and *invention* (which relies on calculation; an engineering model) has also been blurred if not entirely clouded over in this era where a plethora of consumer 'choices' passes for personal freedom. The computer itself has become a voracious 'monomedium', swallowing up all the other media, most of the arts, and even a great many manual crafts into the mere pushing of buttons. It has also demonstrated an unsettling power of mimicry, the ability to simulate anything and everything, so that human powers of discernment are pushed to the limit trying simply to distinguish between the *real* and the *unreal* (nowadays, as Baudrillard phrased it, "more real than real"). In this shift to a 'post-modern' worldview, all boundaries break down, all distinctions become indistinct, and new esoteric mathematical theories of chaos and complexity have surfaced to explain our increasingly fractured 'world picture'. But most people are still in the dark about what it all means.

Just this today becomes the paramount question. However powerful these applications of information theory over the past half-century, they rely on a primitive and reductionist model of human communication which concerns itself solely with the *quantities* of information transferred from one electronic terminal to another, while completely ignoring the qualitative aspects of human understanding and interpretation. Again and again, the question of *meaning* is deferred in favour of refinements of the *medium*, the vehicle for transporting it. (Nobody will pay for 'content' on the Net, only for 'access'.) Indeed, the society at large has been so fixated on the myriad applications of info-tech that the question of the meaning of a human life lived within its automated parameters has been only meekly raised.

Yet the arts of cultural interpretation have hardly stood still while the world has been turned into a gallery of 'bits and bytes' ruled by a faceless regime of transnational gatekeepers and digital domain controllers. Over the past half-century, the patently Western assumptions of this information processing system, and above all its claim to universal validity, have been summarily called to account. In the philosophy of language, in post-colonial literature, cross-cultural hermeneutics and inter-religious dialogue, new depths have opened up which radically counter the trend to bland superficiality in the global marketplace. Figures like Heidegger, Gadamer, Mumford, Ellul, Chomsky, Illich, Gutierrez, Eliade, Panikkar, Nishitani and Fornet-Betancourt will have to stand here for a vast range of thinkers who continue to challenge the cultural monomorphism of so-called 'communications' in the digital age.

Indeed, since the 20th Century also enormously enhanced the prospects for travel, people from many formerly isolated cultures have come into face-to-face contact with one another; the reality of 'the other' proving both more challenging and more liberating than all the images flickering across all the screens. Indigenous alternatives and 'other ways of knowing' have emerged from long quiescence to challenge both traditional Western values and the complacency of this new world information order. Of course those challenging the new configurations of power have been labelled 'anarchists' or even 'terrorists' by the mainstream media (obliged to follow their corporate agendas), but many alternative voices seem to be taking advantage of the Internet and other new media technologies, facilitating communication between people and peoples who would otherwise remain separated. A concerted 'anti-globablisation' movement is gaining momentum all over the globe. The WTO and the G8, let alone the IMF and the World Bank, can scarcely call a meeting these days without calling out squads of police just to keep the people on the streets at bay. This, too, may be considered a matching pair of unanticipated side-effects: on the one hand, the computer's consolidation of business, international finance and government into a single interlocking global 'System'; on the other, the orchestrated opposition to it.

Top of the agenda for the anti-globalisation movement are issues of war and peace, equity and social justice, etc. — issues with a human face which are always raised in concert with increasingly desperate cries to redress the ecological devastation of the planet, which shows no sign of abating despite all the massive and readily available evidence of deforestation, extinction of species, climate change, etc. (News of the day: A recent call to grant the Moon — considered a sacred symbol in every traditional culture — the status of a World Heritage Site is expected to meet with fierce opposition from the USA and Japan, keen to exploit lunar 'resources'.) Today's digital media already testify to the endless human capacity for self-delusion; they so readily disperse attention rather than concentrating it that T.S. Eliot may have written the epitaph for our epoch — "distracted from distraction by distraction." Yet is it clear that they can also function more constructively as a kind of early warning system and global action network. It is still an open question whether these new media will eventually foster genuine human encounter and understanding of any depth or durability. One sign of hope is the growing worldwide opposition to the new arsenal of genetic engineering technologies made possible only by the immense calculating power of the computer and its enormous capacities for data storage and retrieval. Almost totally neglected by scholars of the media, these new techniques for re-ordering the natural world may turn out to be the least expected, and potentially the most dangerous, side-effect of the digital revolution.

Unreality Check — Genetic Engineering

How did it happen? How did we get in half a century from engineering models of machine communication through the set of cybernetic principles governing the computer age to the project currently afoot to re-engineer all of life, including the ominous revival of human eugenics? Here I think Jeremy Rifkin deserves more credit than he generally receives for explicating and critiquing the 'genesis' of this transformation, particularly in the powerful and persuasive chapters "Computing DNA" and "Reinventing Nature" of his 1998 book, *The Biotech Century*, from which I draw in the brief sketch below. Indeed, Rifkin's 'Jeremiad' over the past two decades has not only sounded the public alarm about unwarranted extrapolations of these technologies into every niche of life on this planet, but elucidated for scholars in the humanities and social sciences the recently forged connection between the information sciences and the life sciences. More clearly than anyone else, he demonstrates how 'info-tech' has mutated, almost overnight, into 'bio-tech'.

The information contained in a single biological cell dwarfs by orders of magnitude anything human ingenuity has ever before tried to 'sort and file'. The Human Genome Project alone, the most ambitious data collection project in history, has attempted to catalogue over three billion entries — rather disappointingly, as it turned up barely half as many genes in a human being as in a tulip, or in many of the other species currently being sequenced. This kind of work simply cannot proceed without the computer, indeed, without hundreds of computers in scores of labs around the world, parallel processing, downloading data from one another over the Internet, etc. It should come as no surprise to media scholars that the medium used to investigate the genetic substructures of life has overwhelmingly contributed to the conceptual models and metaphors by which scientists today have come to understand life itself as 'information processing'.

Two potent cybernetics concepts, the 'code' of life (a notion first broached by physicist Erwin Schrödinger in his 1946 book, *What is Life?*) and the mechanism of negative feedback, were picked up by the biologists. First, in 1953 Watson and Crick directly though simplistically adapted Norbert Weiner's terminology to describe the chemical language of DNA as a 'code', an unfortunate term reflecting their earliest gropings in the dark to grasp the 'communication' going on between information articulated in the molecular genetic 'pattern' and the proteins eventually constructed from it. In barely disguised form, the sender/receiver model became Crick's so-called Central Dogma of biotechnology: DNA makes 'messenger RNA' which in turn makes protein. Fifty years on, the scant utility of 'decoding' the Human Genome so far has demonstrated that genes do not 'code' for traits in the one-to-one, point for point fashion implied by this misleading term. (Indeed, the Central Dogma has lately been shown to be as full of holes —or rather as neglectful of wholes: the 'whole' cellular environment, the organism in its entirety, etc. — as the sender/receiver model of communication.[19]) More subtle theorists soon took up Weiner's project to extend negative feedback, the fundamental principle of cybernetics by which a system samples its environment for new information, to living organisms and their evolution. In this view, all organisms adapt to their environment in the same manner as our computers are 'learning' to do. By such a process of reducing complex living systems to information processing systems, evolution has come to be redefined as 'the survival of the best informed'. Around this point, the reductionists rally like a murder of crows to a cornfield. To listen to Richard Dawkins, "If you want to understand life, don't think about vibrant, throbbing gels and oozes, think about information technology." [20] In the flatter language of chaos theoretician Norman Packard, "what drives...evolution is increased computational ability." [21] Or in Freeman Dyson's provocative analogy:

"Hardware processes information; software embodies information. These two components have their exact analogues in the living cell; protein is hardware and nucleic acid is software." [22]

Rifkin comments: "The story of Creation is being retold. This time around, nature is cast in the image of the computer and the language of physics, chemistry, mathematics and the information sciences. With living organisms, as with computers, information capacity and time constraints become the primary considerations. Each succeeding species up the evolutionary chain, like each new generation of computers, is more complex and better adept at processing increased amounts of information in shorter periods of time." [23]

Rifkin's analysis achieves great trenchancy because he recognises the extent to which the 'post-modern' cosmology, the borderless worldview of the information age, provides a legitimising context for the eventual commodifying of all life forms:

"The new cosmology all but eliminates the idea of species integrity. Living things are no longer perceived as birds and bees, foxes and hens, but as bundles of genetic information. All living things are drained of their substance and turned into abstract messages. Life becomes a code to be deciphered. There is no longer any question of sacredness or specialness." [24]

Here, I would suggest, media history may well have an important interdisciplinary role to play. The 50-year-old information processing model underlying genetic engineering seems to be radically at odds not only with the cosmologies of the diverse cultures of the world, but with the traditional value systems of the West. This is not likely to trouble the geneticists, but the fact is that this engineering model cannot do justice to what is actually being discovered by the bio-sciences themselves. Dialogue between worldviews is crucial once we see science stumbling toward an understanding that our physical being, as well as our social and psychological make-up, consists entirely of the communication of messages, from our genetic substructure right through to our highest cultural and religious values: Here the 'word becomes flesh' quite literally and demonstrably. This is hardly news to some of our most venerable cultural traditions, though it certainly adds new evidence, new cogency, and new relevance to their instincts. Quantum physics taught us that we live in a world of energy in constant transaction and transformation. Now modern bio-science seems to be on the verge of comprehending that these communicable patterns of energy may also be *meaningful* relationships. Hans-Georg Gadamer's celebrated hermeneutic formulation, "Being that can be understood is language,"[25] turns out to be far more inclusive than the deterministic credo that your love-life is coded in your genes. As we have seen, the positivistic philosophy of language 'informing' information theory is a crude reductionism meant to deal solely with *quantities* of data — surely unsuitable for the mature discourse between worldviews required by our globalising technologies, and just as surely misleading as a model of nature's elegant processes. The shift needed today to a new theoretical paradigm practically suggests itself: Instead of a manipulable 'code' of life, *language*; instead of reducing language to one-way (sender/receiver) communication, *dialogue*; and instead of cybernetic feedback loops mechanically 'correcting' life to pre-set goals, a deeper *understanding*. Yet the old paradigm is still strong. The oversimplified, mechanical, quantitative system we have been at pains to critique here remains the dominant model of communication for the sciences, even as they attempt to render the very 'humanity' of human beings redundant in their brave new world. At the least, media history may contribute some critical depth of field to the self-understanding of a society seemingly hellbent on remaking itself, and the entire natural world, in the image of the almighty computer.

The foundational technique of literacy — breaking things down into their simplest components, then recombining them in preferred patterns — has led us to split the atom and now to attempt to recombine DNA into forms of life we consider to be better, or more efficient, or just more profitable. In the nuclei of both atoms and cells, this 'technique' — so valuable for centuries to the literate, historical peoples of the West — has come full circle to threaten the continued viability of life on Earth. Here's an 'unreality' check for you: Are you a) appalled by genetic engineering, or b) can you take it or leave it, or c) do you consider it just one more useful technological augmentation of human powers? Even such a simple 'test' reveals basic attitudes well below the level of conscious decision-making. The fundamental (one might almost say 'fundamentalist') conversion of the language of life into the manipulable signs of a coded 'programme' can only seem normal and 'natural' to a mind steeped from infancy in the conventions of phonetic literacy, with scarcely any other mitigating cultural or religious influence. Otherwise, or to other cultures, genetic engineering can only seem appalling in both its arrogance and its presumption to understand the meaning and manipulate the direction of life by such procedures. Our current global economy is marginalising indigenous peoples all over the globe; the irony is that our own descendants may well look back upon us as the last indigenous people, that is, the last generation of genetically unaltered people to know what nature was like before it was all turned into a batch of bioplasmic commodities for global markets. Once we start eating genetically modified food (never before consumed by humans; who knows what will happen?), once we start hybridising human/ animal chimeras across species boundaries, once we start making irreversible 'germline' interventions in human biology, our very bodies will have become 'the medium', in a perverse sense, for the ultimate triumph of cybernetics over biology, of automation over humanisation, and of mere technical 'cleverness' over all the great cultural traditions of wisdom. The pioneering humanist/chemist Erwin Chargaff, whose discovery of base-pairing ("Chargaff's Rules") helped make genetic engineering possible, was among the first to sound the warning:

"I consider the attempt to interfere with the homeostasis of nature an unthinkable crime. Have they peeped into the Creation and found it wanting? We do not yet have a pathology of scientific imagination; but the urge to change the biosphere irreversibly could make an excellent object for such a study." [26]

Recently, Chargaff's misgivings have been echoed by a growing chorus. Andrew Kimbrell, Director of the International Centre for Technology Assessment, writes: "The scientific elite has come to realise, albeit slowly, that current technology is not compatible with the sustaining of life forms. Their solution, however, is not to change technology so that it better fits with the needs of living things, but rather to alter and engineer life so that it can survive and become more compatible with the technological milieu... Through biotechnology life is being absorbed into technology, both at the conceptual and the genetic level... Given this prospect, it is ever more urgent that we become heretics to the religion of science and that we reinvent and devolve our technology. Failure to do this will forever bar a rapprochement with nature, for nature as we know it will cease to exist." [27] It gives one pause, does it not? And it may sound the proper note on which to conclude this survey of media effects.

As we have seen, the great lesson of media history is that the side-effects of new media technologies are not only vast and various, but extremely difficult to foresee. Our ever-accelerating technological society resists looking over its collective shoulder; we're much too busy racing into the future. Yet our view of the future is always necessarily a narrow one, projected through the telescopic lens of the present, while only hindsight offers a sufficiently wide-angled scope to see the side-effects in proper proportion. It took well over a thousand years for the innovation of literacy to eventuate in classical civilisation, and several hundred beyond that for the copying of manuscripts to come to fruition in the Middle Ages and the Renaissance. It is nearly 555 years since Gutenberg's invention of the printing press, and we are only now getting a handle on the transformations of society inaugurated by it. A hundred years were needed for Morse's transmission of words by the medium of electricity to be formalised in cybernetics and information theory, and a further five decades to build up from those principles the information society we so pride ourselves upon today. These things take a while to work themselves out, and even longer to see properly and assess for what they are.

Suddenly we have the ability to break all of nature, including human nature, down into its simplest genetic components, and recombine these into what we suppose may be 'better' forms of life. What will be the side-effects of *that*? All we can conceivably know before the fact is that tinkering with the genetic language of life as if it were a mere engineering 'programme' will have effects which are literally incalculable, and physically irreversible. Do we know enough? Do we understand the ramifications of our actions well enough to undertake such a project on a planet-wide scale?

I don't know. I seriously doubt whether anybody *can* know what the outcome will be. But the entire history of media 'effects' suggests to me that it's high time to pause here for reflection.

An Intercultural Corrective —
The Ideogram as World Picture

What if we looked at all this from another angle? What if we were able to peer at the Western 'engineering model' of life through the semantic web of another, non-alphabetic culture? What if instead of accepting that the 'post-modern' worldview of modern bio-science will 'inevitably' efface the natural world along with all earlier value systems of the West, we were able to find alternatives and the possibility of cross-fertilisation in another culture's understanding of language and communication? The possibility alone would at least be worth exploring. If the technique of genetic engineering derives ultimately from the side-effects of phonetic literacy, print and the digital revolution — all accepted elements in the 'progress' of Western civilisation — then simply as a point of comparison one might reasonably turn to a non-alphabetic culture. Many indigenous protests have been raised to the application of these techniques; these tend to be routinely dismissed as 'unscientific', or 'primitive', or even 'fundamentalist'. What we need to 'weigh in' as an alternative to the alphabetic mentality is an equally literate, but non-alphabetic worldview. The language and traditional culture of China would seem to be the obvious place to turn. Here is a continuous cultural tradition older than that of the West, literate for as long as the West, and yet one which does not share the structures and inevitable strictures of the alphabetic mentality. Fortunately, we may take a shortcut here by following in the seven-league footsteps of more than one remarkable pioneer in the encounter of China and the West. We shall have to look at the Chinese language through the eyes of a poet, rather than through the more narrowly focused lenses of science or history, but this cross-cultural perspective may help us discern the root error which has led the West into this particular *cul-de-sac*.

Ezra Pound in Rapallo, Italy, 1928.
Cover illustration of 'Pavannes & Divigations'.

1 The Linguistic Challenge:
A Poet's Eye View of the Chinese Written Character

Respect for the kind of intelligence
that enables grass seed to grow grass;
the first cherry stone to make cherries (Con193)

It was from Italy in the 1930s that Ezra Pound began to translate
the Stone Books, the Chinese classics carved on giant stone tablets
and preserved to this day in Beijing's Confucian Shrine. Pound's
involvement with China had already been long and deep, running
from *Cathay*, his early Li Bo (Rihaku) translations, and his editing
of Ernest Fenollosa's *The Chinese Written Character as a Medium
for Poetry*, all the way through his own *Cantos* to these lucid, if not
always literally accurate, translations of three of the Four Books,
and later an ambitious attempt to render the Classic Anthology itself
(*Shi-Ching*). At a 1999 conference on Pound's work in Beijing, Kim
Jong-Gil of Korea, the senior Asian scholar/poet present, pronounced
Pound "the only major Confucian poet of the 20th Century" —
Eastern or Western. His pivotal role in the encounter of Western
and Chinese traditions of literacy is by now well established. There
is no need to rehearse here all the old battles about Pound's politics
or difficult personality, which I have taken up elsewhere.[28] Suffice
to say that the ideograms of these old Stone Books spoke to Pound
with surprising vitality, and may turn out to have something crucial
to say to us today about the attempt to 'decode' life as if the whole
of its meaning and value came to no more than a 'code' or an
engineering 'programme'. The quantitative 'sign' language of
information processing faces us with the linguistic challenge to
broaden and deepen our understanding not only of how our own
alphabetic culture got to be the way it is, but how other cultures
might have organised themselves on other premises to different ends.

The original idea of the characters, traditionally ascribed to Fu-
hsi, was of course pictographic, and can clearly be traced back as

far as the 18th Century BC on the great bronze Shang vases. The oldest scripts, called *guwen*, and still largely pictographic, were fixed in standard shapes by the grand recorder Chou in 800 BC. These *dazhuan* or 'greater seal' characters were also called *kedouzi* or 'tadpoles', indirectly due to Confucius' evident interest in them. Weiger relates that in the second century, when the house where Confucius had lived was being pulled down, a cache of old books written in the ancient characters was discovered in a hiding place. Kung, the prince of Lu — not a learned man, remarks Weiger — seeing the big heads and slender tails of the old writing, declared: "These are tadpoles," and the name stuck.[29]

What should stick with us is that t   nturies after his death, these old books with their funny pict   characters appear to have been the most prized possession o.   'ly of Confucius. In the ensuing centuries, knowledge of the ol   cters declined — scholars will recall Confucius' complaint tha. ... orians were no longer leaving blanks for the ones they didn't kno ʾ — and new ones were invented, sometimes cleverly, sometimes clumsily, not mainly by the creation of new primitives, but by combining pre-existing elements. The later ideograms rely heavily on these phonetic elements to convey the sound, while the pictographic character of the radicals — the original elements of meaning — has subsequently been obscured.[30]

For those of us who are not scholars of Chinese, there is a simple point here. The origins of Chinese writing are pictographic, not phonetic. The phonetic alphabet is, as Walter Ong reminded us earlier, the foundational technology of the West.[31] A living organic whole — the spoken word — is broken down into meaningless phonetic units (letters), and these are reassembled into preferred patterns from which a simulacrum of speech may be reconstituted. As we saw in detail above, the main effects of any media technology, literacy included, are simple enough to outline, but its side-effects can be diverse and deep. To recapitulate: In the case of literacy, the side-effects amount to little less than civilisation itself (city-culture, centralised authority, money, numeracy, analytic thinking, causes and effects, lineal time, estrangement from nature, etc.)[32] In the further refinement of print technology, as McLuhan liked to point out, the side-effects of hypertrophied phonetic literacy range from nationalism and religious reformation to scientific method and the division of labour common to all forms of industrialisation, as well as to mass production and, indeed, all the trappings of the modern money economy.[33] Nowadays phonetic language is being digitised and sent round the world at the speed of light, and we are barely even beginning to assess its powerful social, environmental, psychological and religious side-effects (global economy, so-called 'development', environmental degradation, political inter-dependence, the monoculturalism of Hollywood, etc., not to mention genetic engineering).[34]

Returning to the primitive ideograms, one immediately sees the difference. As I am not a native speaker, nor even a scholar of Chinese, I can only testify as so many have before me to the impression ideograms make upon one coming to the language from the outside. The impression of the simple radicals and primitives is that of a direct connection with natural forms. In phonetic languages, that creature depicted in the four-legged ideogram *ma* 馬 would be called a horse, or *un cheval*, or *ein pferd* (to revisit Saussure's famous example of linguistic arbitrariness). In Western languages, the sound is clearly registered by the letters of the alphabet, but there is usually no discernible connection of that sound with the animal. In the ideogram, by contrast, the four legs and tail are *there*, you can practically see the horse, but the connection is lacking between the picture and the sound you are to make. You have to memorise those. As a matter of fact, in different tones *ma* is 'heard' as another word entirely, meaning strikingly different things, 'mother' and 'hemp' among them. Due to this hiatus between the written word and its sound, Chinese dialects have been free to diverge widely, and the spoken language to be radically simplified, while the writing remains legible across all the dialects. Even writing from the time of Confucius is closer to the modern reader than Chaucer is to us in English.

A simple example may clarify what I'm getting at. Let us just count to three in Chinese and look at the primitives underlying these basic numerals:

一 *Yi*, the number one, or unity, representing in compounds the sky, or a roof, or any covering. Cf. *Tian*, 天 Heaven, the vast.

二 *Er*, the number two, or duality, represents the Earth, making a pair with Heaven. It figures in *ren*, 仁 the fundamental Confucian virtue linking Man with his neighbour, benevolence, mutuality; also perhaps, as in Pound's surmise, "the man and his full contents." (Con 22)

三 *San*, the number three, is also the number of Heaven and Earth with humanity in between; the three worlds of practically every traditional religion: "Heaven, earth, man, our law as written/ not outside their natural colour," as Pound would have it (99/698).

So if we want to see what a King is in the ancient Chinese conception, we must first look to the ideogram:

王 *Wang*, King, "the one I," says Weiger, "who connects together Heaven, Earth and humanity." The ideogram itself portrays the King on — or rather *as* — a shamanic *axis mundi* between these three worlds, and thus depicted one does not easily forget the rites the King must perform in order to secure their blessings for the people: "He who understands the meaning of the rites to Heaven and Earth will govern a kingdom as if he held it lit up in the palm of his hand." (Con 147)

Pound himself did not begin his systematic study of Chinese characters until the mid-1930s, when he launched off on the Chinese history Cantos. He came, as I come today, from the outside; he had no formal training in Chinese, nor any contact with the language as spoken. So he is judged inadequately prepared, and his "invention of China"[35] from *Cathay* all the way to the final *Cantos* is faulted accordingly. Following Fenollosa's lead, and therefore indirectly Emerson's ("Language is fossil poetry"[36]), Pound takes up the primal 'speaking' of the elements themselves. He teaches himself Chinese by direct attention. He lets the language itself instruct him by attending to its signal difference from phonetic writing, the pictographic elements. I think what motivates Pound is not so much a romantic quest for an idealised 'language of nature'[37] as it is a recognition that the 'nature of language' amounts to a good deal more than simple decoding. We are losing sight of this due to the influence of what Martin Heidegger once described as "today's pervasive linguistic science which, by turning language into a servant of the technological world of the computer, in fact furthers the destruction of language." [38] What letters we use or what noises we make in any given language may be more or less arbitrary or conventional; that we must speak in order to be who we are is not, and never has been, merely a matter of conscious choice: "There is a must at the root of it." (99/709)

Stone tablets bearing the full texts of 13 Confucian classics

2 The Philosophical Challenge: Symbol ≠ Sign

Encountering another culture, one might expect to encounter other attitudes and approaches to language. From Plato through Descartes and Hegel, Western culture has believed that conceptual thought has direct access to Being. It is one of the pillars of Western philosophy, even though such an assumption is actually counter-intuitive. Our normal experience is that thinking follows being. You have an experience, and then you reflect on it. Your reflection 'catches' the experience, discovers and even crystallises the experience into a concept, but the experience is primary. And yet Western science, classical as well as modern — and emphatically so in the post-modern science of genetic engineering — is built upon the premise that nature can be made malleable to the dictates of reason, that life can proceed from logic, from calculation, indeed, from mere programming. Pound's encounter with Chinese ideograms, by contrast, led him to discover that language may indeed have a more primal relationship to 'things' than do concepts.[39] Human experience is deeply vested in language; we might say that our experiences arrive clothed in language, and can never be entirely disrobed without vanishing utterly. To invert this normal relationship between thinking and being, therefore, requires bending backwards the linguistic joint between experience and thinking: The polyvalent words in and through which our experience expresses itself have to be reduced to terms with a single unequivocal meaning, multiplex traditional symbols must be dismissed as arbitrary signs, and language as a whole squeezed into the narrow straits of a calculable 'code'.

The theory of language as convention (Plato, *Cratylus*), stemming from phonetic literacy itself, stated forthrightly by Locke,[40] and resurrected in the contemporary semiotic doctrine of the arbitrary sign (Saussure, Barthes, *et al.*[41]), turns language into merely a tool, into what is 'useful', or expedient.[42] If meaning is only an arbitrary construct, then what this word — or any word — means is what I darn well say it means. Seriously: Power is what 'arbitrates' the arbitrary sign; power is the real symbol which speaks in the doctrine of the arbitrary sign. It is the rule of Force, rather than Form, John Ruskin might say here.[43] I hit you over the head with my 'word', or rather with my 'terms'.[44] If I have the power, you are compelled to accept the meaning I give those terms. If you are stronger, or one day acquire the power, you overturn my meaning.[45] Instead, Pound found in the doctrine of *zheng ming* (*cheng ming* in his writings) the intuition that cooperative humane community has its 'formal' basis in language. As he demonstrates time and again in his own verse, the 'word' (*logos*) is dialogical (*dia-logos*), the giving and taking of meaning, not my 'word' (my monologue) versus yours. Indeed, the only direct and undisputed words of Confucius the tradition possesses are his *Dialogues*, the conversations with his disciples rather obscurely known in English as the *Analects* (*Lun Yü*). The making and — maybe more basically — the finding of meaning must therefore be seen as a cooperative act, not only as it were horizontally, between human beings in their various social conventions, but also so to speak vertically, making meaningful connections on every tier of the real from the cosmic to the divine.

The 'decoding' of Life into merely arbitrary 'signs' blocks the direct and spontaneous revelation of all Life as symbol, as it once revealed itself to Emerson: "Things admit of being used as symbols because Nature is a symbol, in the whole and in every part." [46] The artificial 'sign' intervenes in the normal and natural relationship between human life and the life of the natural world, so much so that their connection — deemed sacred in practically every longstanding human tradition — is no longer recognised, or even experienced. Herbert Fingarette rightly insists that the distinctive contribution of the Confucian tradition is precisely its recognition of "the secular as sacred." [47] Much as the Daoists attended to the patterns 'already there' in the "Uncarved Block" of Nature, the Confucians paid scrupulous attention to the homely patterns of the everyday amenities which make human life social, and social life humane — they hallow the greeting of strangers, the mannerly relations of family members, even the sharing of a pot of tea. In his *Science and Civilisation in China*, Needham tells us that *li* — which eventually names the entire Confucian 'ritual' organisation of human behaviour,[48] — is "in its most ancient meaning... the pattern in things, the markings of jade or fibers in muscle."[49] Indeed, 2,500 years of continuous Confucian tradition tell us that in the formation of human communities, a respect for this "law of form," this underlying "pattern in things," has a decidedly more humane outcome than relying on the "law of force." In Pound's shorthand: "Peace comes of good manners" (99/698). The powers of *li* are vast and seemingly magical, acting as they do over time and distance without the use of force: "Spontaneous coordination rooted in reverent dignity," as Fingarette summarises 'the rites'.[50] Participating in this net of ritual relationships is thus nothing mechanical, as the Confucian politeness has so often been caricatured, but a spontaneously coordinated collaborative effort — a form of dialogue not only with peers and contemporaries, but also with ancestors and, ultimately, with the conjoined powers of Heaven and Earth. "Human life in its entirety finally appears as one, vast, spontaneous and holy Rite: the community of Man."[51]

Language here is therefore not just a tool or an arbitrary social construct (although it *can* be used as a tool and is *also* socially inflected, as it has been in the West by the letters of the alphabet), but rather *legein* is *phanein*, telling is showing, and also vice-versa; *logos* is *eidos*, the word is image, and the image word or, if you like, revelation.[52] This is not solely a Greek or a Christian intuition, but other such sacred 'words' — the feminine *vac* in the Hindu *shruti*, for instance,[53] or in this case the *zheng ming* of the Neo-Confucians — surely speak in and for their own contexts and make their own distinctive 'revelations'. Even in the West, the *logos* is until the modern age far more than just a code. The code itself begins not as a device for spies, but as a *codex*, a book, back in the days when the primary book was still Nature herself: "In the beginning was the Book," says Norman O. Brown,[54] recalling not only the medieval Book of Nature, still alive in Shakespeare's "tongues in trees, books in the running brook,"[55] but the 'speaking' of those things themselves in meaningful images and integral patterns of relationship.[56] Pound's Chinese studies, his introduction to the notions of *li* and *zheng meng* and the way he correlates these with his own deep reading of the

Western *logos*, seems to have inoculated him against "the discouraging doctrine of chances," (P/202) i.e., in this case the doctrine of language as a collection of arbitrary signs:

> The sages of Han had a saying:
>> Manners are from earth and from water
>> They arise out of hills and streams
>> The spirit of air is of the country
>>> Men's manners cannot be one
>>> (same, identical)
>
> Kung said: are classic of heaven,
> They bind thru the earth
>>> and flow
> With recurrence   (99/698)

    Pound, one should add, was hardly blind to social constructions. What he did see as an arbitrary social construct was *money*, which he stoutly refused to take as an ultimate symbol or measure of value. As a matter of fact his economic reforms — he applauded Kublai Khan's introduction of paper money as a step in the right direction — may well have been thwarted simply by people's persistent tendency to worship the stuff. (Contrariwise, those who devalue all language as an arbitrary semiotic currency sometimes seem to ignore this aspect of the economic regimes conveniently supporting their own 'positions'.) Politics aside, however, if we follow Pound's lead we learn again to read in that Book of Nature "the way things are," instead of presuming they can only be the way we think they are or, indeed, compel them to be. Life is its own spontaneous revelation, and this is a 'gift economy'. The meaning is in the giving and receiving of Life, i.e., of the Word, the *logos*, the principle of intelligibility, the very matrix not only of our social 'constructions' but of our awareness of ourselves dwelling as humane beings somewhere between the vast of the Sky and the deeps of the Earth.

    The hinge between Being and Thinking is *language*. Language as symbol is neither wholly an 'object' in the world, nor solely a construct in the mind of the 'subject'. It is the link, the bridge, between mind and nature; its 'verbal' and metaphorical character, which Fenollosa emphasised, at once differentiates what it unites, and unites what it distinguishes. Language as symbol, as resonant Word, puts us in an entirely different world from that of language as rational calculation, programme, or arbitrary sign. In the latter instance, the world amounts to no more than what we humans say it is or turn it into; only those 'traits' we choose may 'express' themselves, or so we suppose. In the former, things speak for themselves. Phenomena are as Pound calls them *monumenta* (from the Latin *monere*, its root *mens*, mind), they 're-mind' you, call for your attention, cause you to think... The rumble of thunder speaks volumes, the wind and the rain sigh and sing, "Trees die and the dream remains" (90/609), "the rocks/ the bare rocks/ speak!",[57] and people, poets especially, learn how to speak by first looking and listening:

> "We have," said Mencius, "but phenomena."
> monumenta. In nature are signatures
> needing no verbal tradition,
> Oak leaf never plane leaf.   (87/573)

Never, that is, until somebody manages to cook up a transgenic 'oakamore' tree in their biotech laboratory.

This prospect of genetic engineering arises today only because that primal relationship has been so long and violently inverted. The natural vector from Being through Wording to Thinking has become Thinking (reduced to the reasoning Reason) twisted into Programming (or Planning, or Calculation) and then forced into Being (or some 'virtual' Semblance thereof). The living *logos* must first become mere logic if it is to be deployed as logistics... military, economic, scientific and commercial strategies for conquest and control. If the *sym*-bolic is etymologically what 'throws together' the human subject and the natural object, then what throws them apart may indeed be *dia*-bolic, as we earlier heard Miriam MacGillis asserting: "Genetic engineering is evil." But left to our own devices, those of us raised in a world of phonetic literacy all too easily tend to see the entire world as a perfectible text, and therefore to accept such manipulations as inevitable, even 'natural'. Pound's work is not just some antique counterpoint to all that, it is the *antidote*; the discovery of meaning and integrity in 'the Way' things are. We have already encountered Hans-Georg Gadamer's formulation: "Being that can be understood is language."[58] It is not meant as a tautology (Being=Language), but includes the possibility that there may be so to speak a 'face', or 'dimension', or 'urge' of Being not intelligible to us, or manipulable by us. Before his death early in 2002, the centagenarian Gadamer added, "We must think *with* language, not about it or against it."[59]

Allow me to stress this latest perversion of both life and language. To reprogramme evolution, you first have to see evolution — that is to say, the unfolding of life itself — as a programme, which is putting the abstract, conceptual cart well before the living, breathing horse. When Watson and Crick took to calling DNA a *code*, as we remarked earlier, they were consciously borrowing this terminology from Norbert Weiner's cybernetics, where language is almost wholly a matter of coding and decoding, and therefore replicable by machines that sample their environment with negative feedback control circuits.[60] These were chemists thinking like chemists; Watson has long been known for his opposition to vitalism of any sort. The reductionist conversion of words into terms, of a living language into univocal signs — an extension of both the alphabet and the nominalism characteristic of modern science since Occam — is at the roots of genetic manipulation. To seem 'natural', such a procedure requires a mind conditioned from infancy to the conventions of phonetic literacy, breaking down and recombining meaningless elements into preferred patterns. It is a feature of the phonetic alphabet (hardly its best feature), but NOT of the primitive ideograms, where you can still see the man with bulbous head outlined on the bronze vase, or where the ideogram for the sun floats above the horizon line and "they who are skilled in fire shall read Tan 旦 , the dawn" (91/615), and where Heaven, Earth and the humans mid-way between them make a fleeting appearance even in the pictorial elements required to count to three. Pound's Chinese studies stand as a direct reproach to the presumption of biotechnology, and one might predict that his work on the interface between the intelligible 'nature' of language and the figurative

'language' of nature — or "contextual shaping," to borrow Gregory
Bateson's description of biology's molecular 'grammar'[61] — will
grow the more luminous the further and faster we fling ourselves
headlong down the dark path of genetic manipulation.

Consider Ezra Pound's translation of the ideogram *Jing* 句攵
(*ching*4;), customarily rendered simply as 'respect' or 'veneration'.
In his hands, it becomes a one word poem: "respect for the kind of
intelligence which enables grass seed to grow grass; the cherry-
stone to make cherries." The root of the English word 're-spect' is
the Latin *respicere*, 'to look at' or 'look back upon'. By 'looking at'
the pictorial elements in the ideogram, Pound discerns the grass
seed and the cherry-stone and thus the kind of creative intelligence,
eminently worthy of respect, which produces and preserves the
unique character of each. Today it illustrates precisely "the kind of
intelligence" forsaken in genetic engineering.

Pound's respect for the roots of words is therefore a respect for
the roots of life and the depths of human experience. The kind of
attention he paid to the Chinese language offers a sobering contrast
to the current mania, driven blatantly by commercial greed, to enclose
and exploit the genetic commons by manipulating the language of
life itself as if it were a mere code. The most unexpected side-effect
of the information age is not just more cool games and 'sites' to
visit on the Internet, it is biotechnology and the remodelling of Nature
altogether. You may think you've seen it all, but it is not going to
stop with the 'restructuring' of jobs, or universities, or the finances
of once-sovereign nations by today's ubiquitous 'managers'.[62] That's
only where it starts. Genetic recombinations are the most grandiose
management strategy ever conceived: the control of all life — animal,
vegetable and, eventually, human life — by rational programming.
And this is coming to pass not someday in the far future, but here
and now as we day by day find ourselves conditioned to live in ever
more 'artificious' conditions, in a world of "sham things, dummies
of life," as Rilke once called them, produced not by art or creativity,
as is often so grandly claimed, but merely by artifice, clever
engineering techniques, and the 'calculating frame of mind' behind
them.[63]

Yet as the doctrine of *zheng ming* reminded Pound, human words are more than precise scientific 'terms', more than arbitrarily constructed 'signs', or convenient 'signals' for our feelings. The human world of meaning and value is a symbolic universe, as Coleridge and Emerson saw well before Pound, and language is the way we participate in the living reality of that universe. There is a long poetic and philosophical tradition here. The kind of attention Pound — and any poet, really — brings to language is not merely a romantic longing for a 'perfect' primal language (the so-called Adamic language where words are the same as the things they name, with meanings fixed once and for all at the Creation) but a necessary corrective to the debilitating and ultimately self-destructive tendency to level down language to a mere tool for controlling other people, or the natural world. Genetic manipulation is only the most recent (albeit the most grotesque) form of this failure to fathom the reality of human language.

Statue of Confucius in Beijing.

3 The Religious Challenge: Integrity of the Whole

L et me put this argument another way. We know that DNA is in its context *language*, that is, articulated patterns of relationship, though these primal patterns — the famous *signatura rerum* — are presumably geometric at the molecular level, rather than literally pictographic. Suffice to say that by treating this 'language of nature' as a simple code, we arrogantly intervene without ever actually understanding a 'word' it says! And this 'ignorance' goes deeper than the mere five percent of DNA sequences science presently claims to have 'deciphered'. It is as if we were to study a vast, subtle and complex foreign language like Chinese for a week or two, pick up a few nouns and the signs for negation and affirmation, then proceed to delete some and add others in the belief that we are 'fixing' it — with no understanding, without even any attempt to understand what this language of Life might be 'saying'.

So what does it say? For one thing (and perhaps for all things), Nature speaks in *wholes*. This should be obvious. A single cell contains the genetic inheritance of an entire organism, indeed, an entire species. Genetics, by contrast, is a science of bits and pieces; it can make an organism stutter — to express a single trait (red, red, red tomatoes) over and over again; it can force one species to 'speak' a few words in the 'tongue' of a foreign species. (Monsanto's soybeans now 'express' some petunia traits); but the whole will always be more than the sum of such parts. Moreover, the whole 'words' nature speaks always exist in the context of other whole 'texts'. Genetic engineering ignores these larger environmental horizons; its freakish products (potatoes spliced up with toad genes, citrus pumped full of fish antifreeze) take little or no account of whatever other whole systems they may disrupt, from the soil to the farm to the society at large. So whether speakers of Chinese today care to cultivate their own linguistic roots or not — in 1998 Beijing hosted a global conference of 5,000 genetic engineers! — to an English speaker it is just this wholeness of text in context which the early pictograms and radicals reinforce, the very wholeness so routinely and mechanically broken down in the analytic cast of mind fostered by the phonetic alphabet.[64]

Subject and object have been sundered in the West since the beginning of the modern period, as have the secular and the sacred. In his encounter with China, Pound discovers the bridge between these 'poles' of the real. In the Chinese 'three worlds' of Heaven, Man and Earth, Pound finds what may best be called a "sacred secularity", which he then correlates with Dante's 'economic' Trinity in the effort to re-animate ('re-soul') culture in the West.[65] Why is this trinitarian intuition so crucial? Because unity only appears when mapped onto an underlying difference, and difference cannot even be discerned unless there is first 'something' to differentiate. Neither unity (identity) nor duality (difference) is ultimate; neither makes any sense without the other. That first stroke in the primitive 'Yi', ━ , the horizon line dividing Sky and Earth, light and dark, stands not only for the number one and eventually unity, but also for this first 'difference' between above and below, the syndetic axis of the symbolic world upon which, the History of Religions tells us[66], all later differentiations of time, space and human relationships will be overlaid. All of this — the entire symbolic complex of meaning and value we call language — follows from that single stroke, at once distinguishing and binding together the three worlds of practically every traditional culture: Heaven, Earth and the 'Middle Kingdom' of human society in between.

The religions of the world range in their response to the ultimate from a transcendence so transcendent that to affirm or deny it becomes a contradiction in terms, as in the Western traditions, to an immanence so immanent in the Confucian tradition that one can scarcely say anything about it. "Gentlemen from the West: Heaven's process (*Dao*) is quite coherent, and its main points perfectly clear." (85/552) For putting these primal powers of (alphabetic) abstraction and (ideogrammatic) concreteness face-to-face, Pound deserves his place as a genuine pioneer in the encounter of religions. He does not simply collect "The Sayings of Confucius," as Thoreau and Emerson did in *The Dial* a century before him, but judiciously counterposes the immanent immanence of China to the transcendent transcendence of the three monotheisms of the West. This is a breakthrough. In Pound's hands, Confucius comes to life as light philosopher not in some naive attempt at historical representation, but as a living mind, an unshakeable integrity, whose 'Sky', *Tian*, is "perfectly clear" and cloudless, without metaphysical props, and therefore throws a brilliant side-light into the so-called 'negative' theology — with its "superabundant ray of divine darkness" — upon which the nine tiers of Dante's *Paradiso* and the celestial hierarchies of Eriugena and the Pseudo-Dionysius depend. The very juxtaposition corrects the sometimes excessive 'otherworldliness' of the Neo-Platonics with the down-to-Earth Confucian virtues, while offering the Confucian tradition "what is usually supposed not to exist" (Con 95), namely a metaphysical dimension: the clear light. Light radiating in its own space is invisible; it can only be seen at the moment it strikes something *else*. So both traditions may grow new branches, without being uprooted from their native soils. Each begins to shed light on, in and through the other. Transcendence and immanence meet in the endless resourcefulness of that source, at once luminous and numinous... "To see the light pour in, that is, toward sinceritas." (99/694) Pound's work puts in our hands the ideogram for the sun and moon standing together, the total light process of East and West 明 . "The blossoms of the apricot/ blow from the east to the west/ and I have tried to keep them from falling." (13/60)

"Poets translate," Hugh Kenner writes of Pound, "to get into the language something that was not there before, some new possibility."[67] As a footnote here, but certainly a main text elsewhere, I can tell you that the work of Raimon Panikkar stands as half a century of consistent testimony that the 'trinity' Pound found in the Chinese classics — of Heaven, Earth and Man; or *theos, cosmos,* and *anthropos* in Panikkar's terminology, the divine, the earthly and the human[68] — is not the exclusive possession of any single tradition, but may well be the key to intercultural understanding altogether. Why? Any honest translator knows the answer... Because nobody can properly claim to have access to the whole of human experience, let alone the whole of reality. Why not? Because reality is inherently relational, that is, neither an undifferentiated unity nor an unrelated plurality of monads; because there is a built-in polarity, relatedness, interdependence of each with all and all in each. What does this tell us?

In the case of China and the West, as Pound saw, it tells us that we who have grown accustomed to either alphabetic or non-alphabetic worldviews need each other and, ultimately, need as well the humility — which may be another way of saying the courage — to acknowledge that there is between us something which surpasses us, that mystery we call life or light or the divine or the *Dao* of Heaven. The wholeness of the ideogram presents itself as a crucial intercultural corrective to the broken bits of the Western alphabetic mentality.

In the case of genetic engineering, it should be clear by now that wherever worldviews collide, there also arises the possibility of cross-fertilisation and even collaboration. We began with the challenge of genetic engineering to traditional worldviews. We see now that the dialogue of cultures will equally present challenges *to* the culture and mindset which condones genetic engineering. Simply to ask what it means to be human reminds us that no single culture's worldview has the whole answer. These new techniques for manipulating life have put us all on the edge of a precipice: Is this the End of Nature? We can see no bottom to the abyss before us. But if we call out for a little help, instead of blindly plunging over this cliff, if we begin to look and listen carefully — not only to the latest 'breakthroughs' and predictions of modern science, but to all the other human ways of knowing and being which today must enter the dialogue about genetic engineering — then we might just hear echoes from distant shores, and catch sight of one another as human beings, each building our own sorts of bridges to span the immeasurable gulf between today and tomorrow.

The Biotech Time-Bomb —
A *Pluralistic Dilemma*

The Choice

The Ecologist calls genetic engineering the "most pressing environmental issue" of our times. With all the disasters on their plate — from pollution to hunger to global climate change — why should such an influential environmental journal highlight this one issue as the most pressing? I think it is simply because decisions we make today on genetic engineering may well have irreversible repercussions for the future of all life on Earth. Many of these other problems may be set right, or halfway right; but once the genetic genie is well and truly out of the bottle (or the lab), there's no putting it back. Yet, this most pressing issue is difficult for ordinary people to evaluate or even to take seriously because its effects are by and large invisible without a microscope, at least for the moment. If you explode an atomic bomb over a city like Hiroshima, the immediate catastrophic effects give way slowly to residual radiation effects on future generations. The effects, although drastic and terrible, are gradually lessened with the passage of time. But the release of genetically modified organisms into the environment will follow exactly the reverse temporal trajectory. The immediate effects will be practically invisible and seem quite innocuous; yet these effects may well interact and multiply over time in a terrifying geometric progression; they have the potential to be *cumulatively* catastrophic, and to affect all future generations. As we have seen, 'side-effects' have a way of turning out to be the most profound effects. To put a sharp point on it, genetic engineering is a biological time-bomb, and the fuse is already ticking...

Scholars of technology and human values — most prominently Mumford, Illich and Ellul — have brought to public attention the hidden 'thresholds of reversal' inherent in any technology. During the industrialisation of the 19th Century, these were seen to be the points where further application of industrial processes led to 'diminishing returns'. In the 20th Century, Ivan Illich uncovered moments of saturation when the 'use values' of a particular technology (automobiles, schools, modern medicine) suddenly became 'disvalues'. We earlier met with the idea of negative feedback, which serves as the control mechanism for automated systems. By contrast, the disvalues we must now anticipate may well be the results of *positive* feedback — which means simply 'more and more of the same' — applied to instances of genetic manipulation which at first appear to be entirely successful. The dangers here will not arise from the failures of a given technology, but from its runaway successes. Cloning trees made lots of money for the timber industry, so let's try cloning sheep; okay, that seems to work pretty well (despite premature aging, etc.), so now why not clone a human being...?! Positive feedback sets in when any technology is pursued beyond certain intrinsic limits, as a runaway train or a racing engine without a governor will sooner or later smash itself to smithereens. And these thresholds, where apparent progress 'flips' into catastrophic regress, are notoriously hard to see until it's too late. Already in the first tender years of this 21st Century, we find ourselves confronting several such points of no return at once — in the destruction (so-called 'development') of local cultures and subsistence economies, in the proliferation of weapons of mass-destruction (always, of course, to 'keep the peace'), in the exhaustion of resources (to 'grow' the economy), extinction of species (to make room for yet more human 'progress'), etc.

With regard to genetic engineering, we may already have blithely crossed several such thresholds without even seeing them for what they were, let alone realising what we have done. And all the while, the momentum of positive feedback seems irresistible. In New Zealand, where I live, we probably squandered our best opportunity to halt the spread of genetically modified organisms when the Royal Commission on Genetic Modification issued its "Proceed with Caution" compromise report in mid-2001.[69] This was a policy threshold, a seemingly prudent decision to evaluate each case individually on technical grounds, but in principle an incentive to proceed with a technology which appeared to the technocratic elite to promise economic benefits. (Of course the messy legal question of liability for mishaps, where the true costs come to light, is still up in the air.) Another threshold, a political one, was then crossed in the 2002 elections, when Labour Prime Minister Helen Clark spent her entire campaign for re-election inexplicably bashing the Green Party (her 'natural' coalition partner) for its 'uncompromising' stand against genetic engineering. Why should those Greens be so stubborn and 'unreasonable'? What's wrong with a little compromise for the sake of prosperity? One reason appeared during the course of the election itself, somewhat embarrassingly for Ms. Clark, when journalist Nicky Hager revealed that genetically modified corn may *already* have been released on her watch in New Zealand.[70] I shall not debate the merits of this particular case, except to note that such contamination would suit the biotech companies very well, since there would no longer be any 'organic' future for New Zealand's 'greenies' to defend. An important environmental threshold would have been breached without any public or political decision whatsoever. (A recent report by Britain's Soil Association on widespread planting of genetically modified food crops in America found not only that they had not lived up to their promise to increase yields, or decrease pesticide use, but that despite all 'precautions' cross-pollination with natural species was rampant.[71]) In October of 2003, another threshold is looming, this time potentially a legal, political and environmental crisis all in one, with the lifting of New Zealand's moratorium on commercial release of genetically modified organisms. Once this happens — and all the political indications are that it will — the image of New Zealand as a 'clean, green' environment may still be sold on promotional brochures to tourists, but will no longer exist in fact (if it ever really did). New Zealand will simply be one more country in the great global biotech experiment, rather than an alternative to it, or a 'control' for it. People do not yet realise that this is a *totalising* technology, that is, one whose effects are absolute and binding upon *every* living system — biological, cultural and political systems alike. You can't have just a few fields of genetically engineered corn: It's all or nothing. Consider the power of the west wind that sweeps relentlessly across

New Zealand every spring, or ask yourself who is going to tell the bees not to cross-pollinate the flowers? Moreover, once such technologies begin to alter the natural world, they also inevitably show their totalitarian tendencies in the political realm as well (just look at the history of the nuclear power industry).

Now I have called in this book for dialogue on the issue. That sounds nice. Everybody should have a talk about it, both sides can give in a little, and we'll all go home happy. Really? What can dialogue possibly mean with respect to such a technology, and in such a charged commercial and political environment? Can there be a compromise with such an uncompromising, unforgiving technology? Where is the middle ground?

To focus the issue sharply: We are today, one and all, willy-nilly, like it or not, in the midst of making a choice — one I do not believe should be left up to the technocrats and experts and politicians alone — a choice ultimately between two kinds of world. New Zealand is, as noted earlier, a microcosm which includes a fair sampling of all the world's climates and environments. What happens here mirrors what is happening, or may happen, to the Earth as a whole. The easy option is to choose conformity. Yet we might instead choose to set an example to the rest of the world, as we did in the eighties with the nuclear issue. But New Zealanders first have to realise what the stakes are. The stark outlines of the dilemma facing us today are as follows:

Either —

a) The entire evolution of life on this planet, as well as the entire history of human cultures, follows a single line of development, so that our present-day scientific knowledge and technocratic 'culture' is its culmination, and therefore the only realistic basis for human life in the future. In this case, we should not mince words. All other cultures are bound to disappear and give way to the modern scientific worldview, and all other forms of life to those either domesticated or engineered or permitted to exist within the utilitarian framework of that worldview. We may proceed 'cautiously', and show some vestigial respect to these 'atavisms', but we should not raise false hopes or foster illusions of 'peaceful coexistence' (for example between traditional Maori values and those of modern science, or between organic farming and industrial agribusiness).

Or —

b) We recognise that the unfolding of life on this planet, and of human cultural creativity, may follow many paths and take unexpected forms which simply cannot be enclosed in the worldview of any single system, let alone in the restrictive worldview of modern techno-science. We must therefore begin to work urgently toward a healthy pluralism of cultures and to restore the threatened biodiversity of the planet. In this case, we should not mince words. All cultures, including modern Western culture, are bound to collapse if they refuse to recognise their own limits. We cannot universalise any single culture's worldview because nobody has a monopoly on being human, and no single culture has the power to reduce either the diversity of cultures to a single form, or the diversity of species on the planet to single-minded utilitarian purposes.

Is there a middle ground? It's a fair question, but the brutally honest answer is "No." On the surface, there can be no compromise with a technology that permits no compromise. There may, however, still be ways to discover some *common* ground...

Poles Apart

B etween these two points of view, there can be no middle ground at the logical or conceptual level. Reason alone cannot resolve a dispute between two incompatible forms of reasoning (each reasonable in its domain), nor can the classical liberal political 'compromise' encompass worldviews so fundamentally divergent that each excludes the other. You may compromise about driving on the left or the right-hand side of the street; you may compromise about whether to use the metric system or the venerable British measures. But you cannot compromise on ultimate issues, when your very life is threatened, for instance, or your children's lives, or that of the planet itself. Such collisions of worldviews are *pluralistic* dilemmas, that is to say, issues irreducible to a single conceptual framework. And yet... there is only one New Zealand, just as there is only one planet, where we all must somehow find a way to live. A pluralistic issue arises not as some theoretical matter for academics, but first and foremost at the level of practical affairs. Says Raimon Pankkar:

"The problem of pluralism arises only when we feel — we suffer — the incompatibility of differing worldviews and are at the same time forced by the praxis of our factual coexistence to seek survival. *How can we deal with incompatible systems?* Today we face pluralism as the very practical question of planetary human coexistence." [72]

Indifference won't do; it is clear that representatives of these divergent worldviews ignore one another at their peril. And, as Panikkar underscores, a 'super-system' is no answer either: "I, the tolerant one, I, the great scientist or philosopher or political leader, I can make room for everybody in my grand scheme of things, and my system will solve all these problems — provided you all stay in the positions I have assigned you." Unacceptable! To pretend otherwise is not only naive, but dangerous. Many of the most intractable issues facing human societies in the 21st Century are pluralistic dilemmas, where we can neither dismiss 'the other' from a unity which somehow encompasses us both, nor reduce them to our own point of view. There is only one Holy Land for Israelis and Palestinians, only one country for Catholics and Protestants in Ireland, only one land for Europeans and native peoples elsewhere (Need we invoke the current 'War on Terror'?) The challenge of genetic engineering is of this type: There is only one planet for the bioengineers and for all the other people and cultures who may wish to pursue the many more traditional ways of life which would be destroyed outright by widespread proliferation of genetically modified organisms. This is why dialogue is as urgent as it is important. Pluralistic issues, as we shall see below, can only be worked through — carefully, delicately, precariously and always provisionally — in dialogue, a form of human relationship which *may* permit (there are no guarantees!) a common ground to emerge, a larger or wider or deeper horizon of meaning, unforeseeable by either party.

Neither I nor anybody else can predict the form this dialogue will take, let alone its outcome. Fortunately, New Zealanders have a deeper experience of intercultural dialogue than most peoples. The Treaty of Waitangi negotiations have taught us over long years the profoundly human need for ongoing dialogue, consultation and collaboration (with all the mutual transformations such a cross-cultural process entails), as well as the inadequacy of quick 'fixes', whether legal, financial or political. If Maori leaders and elders say "No!" to the release of genetically modified organisms outside the lab, as they have done consistently at every possible forum, they should be heard as full treaty partners ('No means No'), not dismissed as 'unscientific' or fobbed off with the Royal Commission's suggestion of a token toothless committee for bioethics and 'cultural concerns'.

Sad to say, genuine dialogue is not how New Zealand has been coping with the issue of genetic engineering to date. The choice our political and scientific leaders have placed before us today is merely an economic debate, where risks and benefits are weighed solely in economic scales. Over the past two decades (as Sharon Beder and others have demonstrated in damming detail[73]), Western societies have permitted a strange reversal in the perceived relation between environment and economy. One's natural expectation would be to see the environment as the larger system, and the economy as a subset of that biosphere, consisting of things people do and make and use for their own purposes. But the upshot of two decades of neo-liberal economic theory has been that environmental decisions are now made strictly on the basis of economic criteria. The environment is thus perceived, bizarrely, as a subset of the economy. This may be due in part to the unfortunate term 'environment' itself, coined in the 1830s by Thomas Carlyle, in which 'Nature' becomes merely a vague backdrop to human affairs. At any rate, changes to industrial regimes, to patterns of consumption, pollution and so on, are undertaken today only if deemed 'economic'. This strange reversal of values pits the marketplace, where individual and corporate greed remains the driving factor, against the common good of the living Earth, with all its interdependent systems and cultures (One example: George W. Bush will not agree to the minimal Kyoto Protocol — little enough to slow the drastic climate change caused by our century-long fossil fuel binge — because he claims it will hurt the US economy. Some people consider this a rational decision.) If we cannot change such basic attitudes, the outcome of the stark choice sketched above may be a foregone conclusion.

First, however, we shall have to overcome the strong impediments to dialogue from *both* sides, from those who speak for the dominant techno-scientific worldview and from those capable of voicing alternatives. "Why should we engage in dialogue," say the powerful, "when we have the power to do as we please?" Indeed! Bleakly summarising the failure of *every* attempt to impose moratoria or security limits on genetic engineering since 1974, technology critic Jacques Ellul writes: "Let us entertain no illusions. Scientists will not accept philosophical, theological or ethical judgements. Science simply leaves by the wayside those scholars who have scruples of conscience. It goes its inexorable way until it produces the final catastrophe." [74] From the other side, the resistance to dialogue is equally fierce. "Why should we engage in dialogue," say the powerless, "when the cards are stacked against us from the outset and our protest is bound to be ignored?" Better to satirise, or maybe subvert or — why not? — sabotage the system, become 'Wild Greens' and pull up engineered potato plants in the dead of night, even if we risk jail and are labelled terrorists. At least we finally get some media coverage and make a statement they can't ignore.

Or do we? Doesn't the latter kind of 'protest' fit exactly the competitive Darwinian worldview where the strong devour the weak, however 'noble' their ideals? One should be clear here. Legitimate protests, conscientious acts of civil disobedience, street marches and rallies and so on are a healthy, vital part of any democratic society. They can also be wonderfully festive occasions, celebrating alternative ways of life. But when things start getting broken, especially when bones and heads start getting broken... you've gone too far, and set yourself up to fail. You are no longer an activist, merely a 'reactivist', following an agenda set by your adversary. The system only grows stronger as it easily rolls over such 'disorderly' protests with its legal and police powers. Instead of articulating real alternatives, you risk not only the trumped-up label of 'terrorist', but the danger of becoming what you profess to abhor: single-minded, fanatical, dogmatic, absolutist in your views and destructive in your actions. You risk becoming what you hate, and your 'protest' has the potential to turn almost as nasty as the policies you are protesting. Hence the enduring power of genuinely non-violent activist personalities like Gandhi or Martin Luther King or Dorothy Day, to take only recent examples, and the enduring value of their achievements. There is another way...

So what about the powerful? What do we say to them? Why should they not, like Bush's advisor Condoleezza Rice at the very beginning of his reign, simply announce: "We're the most powerful... we can do what we want" ?! Ah, but beware: This power of yours is deeply ambivalent. In a sense it holds you in its power, and it is so powerful it may end up destroying you, too. Do you see that you cannot *not* use this power, that you cannot refrain from using it or even restrain it to any meaningful degree? He who rides the tiger cannot dismount... there is no space for reflection, you have to keep going faster and faster, yet you have no idea where this power is taking you. The speed of technological change, the focus required to keep on track, the specialised technical know-how deployed, all mitigate against self-knowledge. Here is Ellul again: "The intellectual and cultural tragedy of the modern world is that we are in a technical milieu that does not allow reflection. We cannot look at the past and consider it. We cannot fix on an object and reflect on it. The technical object encompasses us even though we know nothing about it..." [75] Only the 'others' can see that you are trapped in the grip of this irresistible power of yours. Without that 'other' viewpoint, you don't even know what you are doing; indeed, you *cannot* know because you can never stop to consider what you are *un*doing.

Plainly the sender/receiver model of human interaction is very adaptable; it can simplistically 'reduce' practically any human relationship to an either/or polarity. This is a big part of the problem. So far the question of genetic engineering has been reduced to the producer/consumer relationship, the same model in another guise. From the one side, we see very powerful transnational corporations, akin and often allied to those which already own most of the media, today claiming that they have the power to cram not only their message but their artificially engineered food down people's throats. From the other side, we have a worldwide consumer rebellion, millions of people saying that they won't be force-fed this way, that they will accept neither the message nor the phony food, but they will — literally or figuratively — vomit it back, reject it at the supermarket or grab their own headlines protesting it. Beyond all the shouting and media static, there's a slippage between worldviews here, a kind of *moral* short-circuit between senders (producers) and receivers (consumers): Even starving Africans reject genetically modified corn dumped on them in 'aid' packages when it won't sell on the American market. If we are to make any headway at all in coming to grips with the deeply human issues raised by genetic manipulation, we are simply going to have to get beyond this crude model of human communications. Ellul again, one last time: "Propaganda ceases where simple dialogue begins." [76]

Openness to dialogue is a fundamental human attitude, not just a means to a predetermined end or some kind of political expedient.

It implies a good deal more than just a conversation you and I could conceivably have one fine day, although that might well be a start. Dialogue has inner as well as outer dimensions and ramifications. It involves opening up a space in oneself for the other's way of life to take root. It stretches the imagination, and deepens one's capacities for sympathy and empathy. It means finding the other 'within' oneself, playing 'host' to the strange notions of the other, taking their words 'to heart' as well as facing them 'head-to-head' across a table in argument. There are risks; you become vulnerable, you may well have to change your mind, or your way of life. Sometimes it is simply impossible: Have you ever seen anyone change their mind in a television debate? — the medium itself (the unblinking spectator's eye) tends to preclude dialogue.

For those like myself, who find the very idea of genetic 'engineering' repugnant, dialogue might require pausing in one's opposition to reflect on the positive possibilities. I'm not a biologist; I have a lot to learn about the scientific study of genetics, about the intriguing search for ancestors in mitochondrial DNA, about the way ancient organisms can be genetically 'typed' to reveal paleolithic environments, about genetic 'fingerprinting' in crime detection, about the many possible salutary medical uses of these new techniques. I would have to concede that medicine has often turned deadly poisons into healing potions. I cannot rule out in advance the possibility that medical research may do so again with genetic engineering, so long as it proceeds with a full measure of respect for the primary Hippocratic dictum: "First do no harm." [77] I would indeed have to admit that whenever I catch a glimpse of what the scientists are actually seeing through their methods and microscopes, etc., I am struck by the sheer *beauty*, as well as the complexity, subtlety and integrity, of the evolutionary processes which have woven the fabric of life on this planet. A poet's eye view of science, to be sure, but I should like to be convinced that the study of genetics can indeed enhance our understanding of life, as well as magnifying our powers to manipulate it.

Contrariwise, for those versed in these sciences, maybe even professionally committed to the project of re-engineering life by these techniques, dialogue may well require something like a 'personal moratorium', a sincere pause for reflection, a deepening of one's appreciation of life, a moment (which might last a lifetime) to cultivate a sense of awe at what one is seeing and doing... a stillness, even a sense of reverence before the 'facts of life', rather than just another round of grant applications. In today's corporate climate, is it no longer humanly possible to formulate — or get funding for! — rigorous programmes of basic research, instead of merely churning out 'living commodities' on demand? If the get-rich-quick financial incentives were removed, how many molecular biologists would go back to their labs simply to *study* — which can be, according to Cicero, "the intense and enthusiastic activity of the spirit directed to something...with all the power of our striving and commitment" [78] — in order to learn what the immense heritage of life treasured up in every living cell has to teach them? Given our current levels of information overload, you can bet it won't be very many years before all the 'sponsored' research we so prize today will be seen for what it is — blatantly biased, almost laughably slanted in favour of the sponsor's viewpoint from the outset — and dismissed as practically worthless. Imagine on the other hand what discoveries, what insights, what startlingly original research these highly trained microbiologists might come up with were they free to explore the basic science of the gene instead of being paid so handsomely to turn life itself into a marketable commodity!

The Gift

Nobody can have a 360-degree 'perspective'; it would be a contradiction in terms. Dialogue can bestow not only the point of view of the other, but something *more*... Something happens, something 'occurs' to both participants that neither could have foreseen or predicted. This spontaneity, this freedom, is what distinguishes a dialogue from a debate or a script or a monologue. It cannot be controlled by rules of the game laid down by either participant. The vision of the whole relationship that emerges in dialogue transcends the participants' point of view, yet both are able to share in it, each according to their lights. Each learns something from the other, and both learn something from the encounter. Dialogue opens a window to the 'Big Picture' neither could see on their own, and opens as well a further vista on the entire range of human experience, i.e., our 'dialogue' not only with other humans but with the whole of life.

A certain type of technocratic mentality will see life as a problem, or rather a series of problems — hunger, disease, suffering, etc. — which can be 'solved' by implementing greater human control over them. Amplified by modern science, this attitude is now producing techniques of genetic engineering that potentially increase the human power to 'control' life to an unprecedented degree. A certain kind of bureaucratic mentality will see these techniques as potentially dangerous or at least upsetting to social, political or environmental order, and therefore seek to 'control' them by laws and 'guidelines' and regulatory mechanisms. The mentality of both groups is strikingly similar; in the final analysis, it comes down to an *urge to control*. For both groups then, the old Roman paradox retains all its force: *Who will control the controllers?*[79] First, we see looming on the horizon genetically engineered organisms which break all the old rules. Second, a new set of rules and regulations must be generated to 'control' these. Next, we'll soon see a new generation of further and more subtly modified organisms that escape these early regulatory restrictions. Then will come more regulations, then more GMOs, and so forth, *ad infinitum*. Even if both the technocracy and the bureaucracy were fully automated, and internally governed by strict negative feedback controls, the relationship between them would still be subject to runaway positive feedback and altogether out of anybody's control!

I am convinced it is time to step back and take a hard look at this paradigm of 'control'. We should be aware that this mentality is neither universal, nor inescapable. It is tied to the development of certain communications technologies — literacy, print, cybernetics — by one particular culture. As we have seen, biotechnology may be considered a vast and sophisticated application of techniques which originally stem from the phonetic alphabet, branch out into the mechanical copying processes of print culture, and are today bearing all sorts of strange 'virtual' fruit thanks to the cybernetic systems governing the 'information revolution'. This in mind, I would like to apply to biotechnology an insight the pioneering biologist/cyberneticist Gregory Bateson first formulated with respect to nuclear weapons and the arms race. In order to do so, I need to delve briefly into the background of Bateson's insight, which involved a conversation — a real dialogue! — with J. Robert Oppenheimer, the man who started the arms race.

The impending terrors of the atomic age were heralded by a seemingly innocuous coded military communiqué proudly announcing the 'successful' detonation of the first atomic bomb (16 July 1945) with the words: "The baby is born." Two years later, the putative 'father' of that atomic bomb, Oppenheimer, put the situation that had arisen after the 'birth' of this new technology to Gregory Bateson in a particularly arresting way. The initial conversation is reported and followed up by Bateson in a most curious 1978 interview with Daniel Goleman.[80] Transcribed below, we may take it as an example of a successful dialogue (between Bateson and Oppenheimer), where a new insight spontaneously 'occurred' to those involved, embedded in another, much later dialogue, where Bateson tries — and largely fails — to convey the import of that earlier conversation to the media interviewer:

Oppenheimer (in 1947): You know, if anyone asked me why I left teaching at Cal Tech and came to do research at Princeton, I suppose the answer was that at Cal Tech there were 500 students to face, who all wanted to know the answers.

Bateson: I suppose the answers to these questions would have been rather bitter.

Oppenheimer: Well, as I see it, the world is moving in the direction of hell, with a high velocity, and perhaps a positive acceleration, and a positive rate of change of acceleration; and the only condition under which it might not reach its destination is that we and the Russians be willing to let it go there.

Bateson (in 1978): Every move we make in fear of the next war in fact hastens it. The old deterrence theory. We arm up to control the Russians, they do the same. Anxiety, in fact, brings about the thing it fears, creates its own disaster.

Goleman: So, just let it happen?

Bateson: Well, be bloody careful about the politics you play to control it. You don't know the total pattern; for all you know, you could create the next horror by trying to fix up a present one.

Goleman: The patterns you talk about in which we are enmeshed seem much larger than we can grasp.

Bateson: There is a larger mind of which the individual mind is only a subsystem. This larger mind is perhaps what some people mean by "God." But it is immanent in the total interconnected social system and includes the planetary ecology.

Goleman: It seems almost futile to try to perceive, let alone control, this larger web of patterns and connections.

Bateson: Trying to perceive them is, I'm sure, worthwhile. I've devoted my whole life to that proposition. Trying to tell other people about them is worthwhile. In a sense, we know it already. At the same time, we don't know. We are terribly full of screaming voices that talk administrative "common sense."

Goleman: Rather than...

Bateson: Wisdom. If there be such a thing.

Wisdom? At once a sense of the limitations of our knowing power, and something else — a sense that there are non-manipulable factors in human life, crucial factors that we simply may not control with impunity. And somehow we do know this, though perhaps we do not know exactly how we know it. We know that we become tangled up in our own machinations, that the very means we employ to control things — or other people — backfire, default on us, escape our control, and end up by controlling the would-be controllers in all sorts of unexpected, very often uncomfortable ways. That wisdom emerged in Bateson's dialogue with Oppenheimer, the striking conclusion of which he seems to have recalled verbatim, but fails to come across in the 'formal' press interview with Goleman. Bateson later put it in a succinct form that I should like to reformulate with respect to biotechnology: *The urge to control is itself the pathology.*

In a nutshell, the only 'control' that will ever actually work on genetically modified organisms is the purely formal one of a 'GE-free New Zealand'.[81] That is to say, New Zealand could have a very important role to play in the great Biotech Experiment of the 21st Century. Our job might well be to 'pass' rather than to 'capitalise' on this technology, to resist all the temptations and the glittering financial inducements, and just let it go... as we have prudently done with nuclear power. Geophysically remote as we are, we still have the option now lost to most of the world to leave this dangerous technology either in the lab or far and away offshore. Otherwise, genetic engineering becomes a planet-wide experiment utterly out of control. There can be no scientific rationale for running such an 'uncontrolled' experiment on the planet as a whole, however rigorously individual projects may be monitored. This said, we should recall that laboratory protocols for these potentially dangerous techniques are very stringent, and should be kept that way. But how can *any* bureaucratic regulatory regime ever 'control' the products of genetic engineering once they're out of the lab? The only 'control' is not to mess with it at all. Hands off! In short, we can still realise the 'deep green dream' of many New Zealanders, Maori and Pakeha alike, and trust that a cleaner, greener New Zealand can instead demonstrate to the world the viability of alternatives — the time-tested organic farming methods, the undisputed nutritional benefits, the promising wind and solar power options, the various alternative medicines, the growing allure of unspoiled bush for ecotourists, etc. — all of which will turn *us* into the 'wave of the future' once the inevitable genetic miscalculations and disasters begin to show themselves.

The real challenge of genetic engineering may simply be *to see life whole*, to affirm life as a whole, with all its terrors as well as its splendours. This by no means implies fatalism, or resignation. Such a vision arises as a spontaneous affirmation. It cannot be forced, or contrived, or faked. It accepts all of life as *pure gift* — knowing that ultimately there may be no giver, and there certainly are no givens. It comes to us, as all good things come to us, *gratis*, for free, and the attitude it confers toward life is accordingly one of heartfelt *gratitude*. The Native American Iroquois Confederacy opens every ceremonial occasion with their famed Thanksgiving Address (ritual basis for the Thanksgiving Holiday in the USA and Canada alike). They do so, they say, because "the thankful mind is the clear mind." That's it! That is precisely the shift in attitude required of us today. In the rediscovery of life itself as gift, indeed as grace, we find the bedrock on which to stand against the threat of total bio-commodification. Without this sense of the sanctity and inviolable integrity of each living thing, and of all living things, there is no hope. Such an attitude receives life, conserves life, and celebrates life, so as to pass it on enhanced to the next generation — *as gift*, not as a pre-programmed, patented 'product' for sale. "THERE IS NO WEALTH BUT LIFE," wrote John Ruskin in capital letters.[82] We live in an age when it seems necessary to stress these things, which have for countless generations been plain and obvious truths. Yet what other purpose has human culture if not to help us recall at least this much about who we are and what we're doing here?

In a letter to the editor of *Science* entitled "On the Dangers of Genetic Meddling", Erwin Chargaff, who contributed as much as anyone else to discovering the structure of DNA, already warned (25 years ago!) that by adding these "freakish forms of life... we shall be throwing a veil of uncertainties over the life of coming generations. Have we the right to counteract, irreversibly, the evolutionary wisdom of millions of years, in order to satisfy the ambition and the curiosity of a few scientists? This world is given to us on loan. We come and we go; and after a time we leave earth and air and waters to others who come after us. My generation, or perhaps the one preceding mine, has been the first to engage, under the leadership of the exact sciences, in a destructive colonial warfare against nature. The future will curse us for it."[83]

Quite recently, and taking aim at this very "warfare against nature," Wendell Berry saw fit to underscore Shakespeare's well-known affirmation that "thy life's a miracle,"[84] by adding his own strong proviso: "To treat life as less than a miracle is to give up on it."[85] Citing a long roster of luminaries who refused to "give up on it," Berry concludes:

"All along, the enterprise of science-industry-and-technology has been accompanied by a tradition of objection. Blake's revulsion at the 'dark, satanic mills' and Wordsworth's perception that 'we murder to dissect' have been handed down through a succession of lives and works, and among the inheritors have been scientists as well as artists. ...Among these mourners have been people of the highest intelligence and education, who were speaking, not from nostalgia or reaction or superstitious dread, but from knowledge, hard thought and the promptings of culture.

"What were they afraid of? What were their 'deep-set repugnances'? What did they mourn? Without exception, I think, what they feared, what they found repugnant was the violation of life by an oversimplifying, feelingless utilitarianism; they feared the destruction of the living integrity of creatures, places, communities, cultures and human souls; they feared the loss of the old prescriptive definition of humankind, according to which we are neither gods nor beasts, though partaking of the nature of both. What they mourned was the progressive death of the earth."[86]

On the brighter side, it's not too late. If we pay attention now, we may not only retain but redouble our sense of the miracle of this life we hold in trust for our descendants. As Jeremy Rifkin has repeatedly asked, "What's the hurry?" Why must we rush and risk spoiling forever this great gift we have been given? The standard retort seems to be that there is money to be made, and if we don't hurry we might miss out... Perhaps every age has to sort out its false gods from its real values. Now it's our turn. Now the time has come for us to discern just what we do value, to remember what really does give meaning to our lives, and maybe at long last,

> Learn of the green world what can be thy place
> In scaled invention or true artistry.[87] (81/521)

Scott Eastham
Waitangi Day, 2003

*respect for the kind of intelligence that enables grass
seed to grow grass; the cherry-stone to make cherries*

Notes

1   Barrington Nevitt, *The Communication Ecology* (Toronto: Butterworths, 1982), iv; reinforcing the earlier formulations by anthropologist Edward T. Hall: "Communication constitutes the core of culture and indeed of life itself."

2   George Steiner, *Errata — an examined life* (New Haven & London: Yale, 1998), 73.

3   Stanley Diamond, *In Search of the Primitive — A Critique of Civilisation* (New Jersey: Transaction Books, 1974), 1. See also the works of Lewis Mumford and Harold Innis along these lines.

4   J. O'Donnell, *Cassiodorus* (Berkeley/Los Angeles: University of California Press, 1979), 234-5.

5   As a matter of fact, the analogy holds right down to the individual letters. The piece of type itself (a 'sort') is a copy of the steel-engraved letter on a punch, long used in making coins and medallions. The traditional view of Gutenberg's innovation is that he came up with the clever hand moulds which fill in with molten metal and stems the 'matrix' made by a punch (or 'patrix'), then assembled the resulting letters into galleys and pages. John Man details the process, as well as alternative theories of what Gutenberg actually achieved, in *The Gutenberg Revolution* (London: Hodder Headline) 2002. According to Man, "What is needed...is a system that reproduces copies of punches... in modern terminology, the information has to be cloned — a perfect copy of an enduring original. It is the copy that will then produce the new medium itself, the printed page" (p. 124).

6   See J.W. von Goethe, *Faust, Part II* (New York: Norton, 1976), Act 2, conclusion of Scene 2.

7   From Ralph Waldo Emerson, "The Adirondacs — A Journal. Dedicated to my Fellow-Travellers in August, 1858", *The Works of Ralph Waldo Emerson* (New York: Tudor, 1930), Volume 4, 247-8.

8   Cited in Tom Standage, *The Victorian Internet* (New York: Walker & Co., 1998), 83.

9   ibid., 164, citing the *National Telegraph Review and Operator's Companion*, 1853

10   H. Marshall McLuhan, *Understanding Media — The Extensions of Man* (New York: McGraw-Hill, 1964), 70-71.

11   Ezra Pound, "The Serious Artist," in *The Literary Essays of Ezra Pound* (London: Faber, 1954; Norfolk, CT: New Directions: 1954), 49. And similarly, in his *Guide to Kulchur* (New York: New Directions, 1952/1970), 77: "The mind making forms can verbally transmit them when the mental voltage is high enough."

12   Derrick de Kerckhove, "On Nuclear Communication," *Diacritics*, Summer 1984, 92-81.

13   Vaclev Havel cited in H. Henderson, *Beyond Globalisation, Shaping the Sustainable Global Economy* (Bloomfield, CT: Kumerian Press, 1999), epigraph.

14   R. Panikkar, "The Destiny of Technological Civilisation: An Ancient Buddhist Legend, *Romavisaya*", in *Alternatives*, Vol. X, No. 2, Fall 1984, 241.

15   C.G. Jung, *The Undiscovered Self*, R.F.C. Hull trans (Boston/ Toronto: Little, Brown, 1957/1958), 99.

16   Jeremy Rifkin, *The Biotech Century* (New York: Tarcher/Putnam, 1998), 184.

17   Barrington Nevitt, *The Communication Ecology*, op. cit., 40-43.

18   George Myerson, *Heidegger, Habermas & the Mobile Phone* (London: Icon, 2001).

19   Fritjof Capra offers a serviceable overview of such current critiques of the Central Dogma in *The Hidden Connections* (New York: Doubleday, 2002), 158-75, drawing on the authoritative works of Mae-Wan Ho, *Genetic Engineering — Dream or Nightmare?* (Bath: Gateway, 1998), and Evelyn Fox Keller, *The Century of the Gene* (Cambridge, MA: Harvard, 2000).

20   Richard Dawkins, *The Blind Watchmaker* (New York: Norton, 1986), 112.

21   Norman Packard, cited in Pierre Grasse, *Evolution of Living Organisms: Evidence for a New Theory of Transformation* (New York: Academic Press, 1977), 226.

22   Freeman Dyson, *The Origins of Life* (Cambridge: Cambridge UP, 1985), 6.

23   Jeremy Rifkin, *The Biotech Century*, op.cit., 211.

24   ibid., 214.

25   Hans-Georg Gadamer, *Truth and Method* (New York: Seabury, 1976), 432.

26   Erwin Chargaff, *Heraclitean Fire* (New York: Rockefeller UP, 1978), 190.

27   Andrew Kimbrell, "Recreating Life in the Image of Technology," *The Ecologist*, Vol. 29, No. 2, May/June 1999, 169-70; see also the entire issue of *The Ecologist*, 28.5 (Sept./Oct. 1998), entitled "The Monsanto Files — Can We Survive Genetic Engineering?"

28   Cf. Scott Eastham, "Modernism Contra Modernity: The 'Case' of Ezra Pound," in *Paideuma*, 30. 1 (Summer 2002), 97-132, or the shorter version in H. Dennis, ed. *Ezra Pound and Poetic Influence* (Amsterdam/Atlanta: Rodopi, 2000), 235-255.

29   I draw this brief history from L. Weiger, *Chinese Characters — Their origin, etymology, history, classification and signification* (New York: Dover/Paragon, 1965), "Historical Sketch," 5-9.

30   Cf. Ping Xu, 'Rose-Cherry-Sunset-Iron-Rust-Flamingo' Diagram and the Genesis of Ezra Pound's Ideogrammatic Method," *Paideuma* 27. 2-3 (Fall/Winter 1998) 53-68, who credibly suggests that Pound's ideogrammatic method may well be less about juxtaposing things than it is about getting to the linguistic roots of our relationships to those very things.

31   Cf. Walter Ong, *Orality and Literacy: The Technologising of the Word* (London: Methuen, 1982).

32   Cf. Ivan Illich & Barry Sanders, *ABC — The Alphabetisation of the Popular Mind* (New York: Vintage/Random House, 1988); Robert Logan, *The Alphabet Effect* (New York: Morrow, 1986).

33   Cf. Marshall McLuhan, *The Gutenberg Galaxy*, (Toronto: University of Toronto Press, 1962); Elisabeth Eisenstein, *The Printing Revolution in Early Modern Europe* (Cambridge: Cambridge UP, 1983); John Man, *The Gutenberg Revolution* op.cit.

34   Cf., e.g., Neil Postman, *Technology — The Surrender of Culture to Technology* (New York: Knopf, 1993).

35   Cf. T.S. Eliot's Introduction to the Faber edition of Pound's *Selected Poems*, (London: Faber & Faber, 1948) 14-15: "...it must

be pointed out that Pound is the inventor of Chinese poetry for our time..."Chinese poetry, as we know it today, is something invented by Ezra Pound," going on from this all too ironic faint praise to disparage the possibility of translation altogether.

36   For the role of Emerson's remark on "fossil poetry" (from "The Poet," 1844) in providing the foundational idea for the project that eventually became the Oxford English Dictionary, cf. Hugh Kenner, *Historical Fictions* (San Francisco: North Point Press, 1990), "The Keys to the Kingdom," 331-338.

37   As Robert Kern suggests in his *Orientalism, Modernism, and the American Poem* (Cambridge: Cambridge UP, 1996). For Kern it is a source of consternation (on p. 7 and throughout) that Jacques Derrida exempts Pound and Fenollosa from his critique of "the European hallucination" of China in *Of Grammatology*, G.C. Spivak, trans. (Baltimore: Johns Hopkins, 1976), 76-92.

38   Martin Heidegger, Letter to "Prof. R. Panikkar und seine Studenten," 18 March 1976, written only a month before his death. Heidegger of course maintained that "in fact it is language that speaks (*die Sprache spricht*). Man begins speaking and man only speaks to the extent that he responds to, that he corresponds with language, and only insofar as he hears language addressing, concurring with him." (George Steiner's translation (epigraph to *After Babel*, (London: Oxford UP, 1975, xi) of the famous passage from "Poetically Man Dwells" included in Martin Heidegger, *Poetry, Language, Thought*, Albert Hofstadter, trans. (New York: Harper & Row, 1975), 215-16. For a fairly elaborate demonstration of how far "today's pervasive linguistic science" has failed to take Heidegger's point 20 years on, one need only turn to A. Lock and C. Heelas, eds. *The Oxford Handbook of Symbolic Knowledge* (Oxford: Clarendon, 1996), 'constructed' entirely on the unquestioned utilitarian assumption that words can be reduced to terms, symbols to arbitrary signs, and all forms of symbolic discourse merely to subsets of tool-use.

39   A similar intuition seems also to be a longstanding tradition in India. Cf. R. Panikkar, *Blessed Simplicity* (New York: Seabury, 1982), "A Note on Thinking and Speaking," 122-3: "In India, the ultimate polarity, the yin/yang so to speak of the Indian effort at

human orientation in reality, is not thinking and being, but *being and wording*. Or rather being and speaking: being and letting being be; being and letting being escape... It is a kind of total spontaneity. Being explodes itself into being, into word, into the expression of that being, into something which goes its own way, like an expanding universe which nothing and nobody — and certainly no being, no thinking, no lack of contradiction, no logic or logistics, no any thing — can control or guide."

40   In his *Essay Concerning Human Understanding*, (A.C. Fraser, ed., (1894); reprint (New York: Dover, 1959) 3.2.1, John Locke declares what has since become the common sense English view, namely that words "came to be made use of by men as the signs of their ideas; not by any natural connection that there is between particular articulate sounds and certain ideas, for then there would be but one language amongst all men; but by a voluntary imposition, whereby such a word is made arbitrarily the mark of such an idea." Here we are following the line of thought first laid down in Plato's *Cratylus* itself; cf. the penetrating discussion of this seminal text in Hans-Georg Gadamer, *Truth and Method* (New York: Seabury, 1976), Third Part, 2, "The Emergence of the Concept of Language in the History of Western Thought," a) Language and *logos*, 366-78: "Greek philosophy more or less began with the insight that a word is only a name, i.e., that it does not represent true being. ...Fundamentally language is taken to be something wholly detached from the considered object and to be, rather, an instrument of subjectivity. This is to follow a path of abstraction at the end of which stands the rational construction of an artificial language. In my view this takes us away from the nature of language." (366/377).

41   Cf., e.g. Ferdinand de Saussure, "The Object of Study" and "Nature of the Linguistic Sign," in his *Course in General Linguistics*, Roy Harris, trans. (La Salle, IL: Open Court, 1983); Roman Jacobson, "Closing Statement: Linguistics and Poetics," in *Style in Language* (Cambridge, MA: M.I.T. Press, 1960); Claude Levi-Strauss, *The Savage Mind* (London: Weidenfeld and Nicolson, Ltd. and Chicago: University of Chicago Press, 1966); Gerard Genette, "Structuralism and Literary Criticism," in *Figures of Literary*

Discourse, Alan Sheridan, trans. (New York: Columbia University Press, 1982); Roland Barthes, *Elements of Semiology* (London: Cape, 1967), etc.

42 Cf. the first three chapters contra the dominion of this instrumental theory of language in Scott Eastham, *The Media Matrix — Deepening the Context of Communication Studies* (New York/ London: UP of America, 1990).

43 Ruskin's "law of help" is applied in his work not only as an aesthetic principle — "Composition may be best defined as the help of everything in the picture by everything else." — but also as the most basic principle of nature and society: "The power which causes the several portions of the plant to help each other, we call life... Intensity of life is also intensity of helpfulness." E.T. Cook & A. Wedderburn, eds., *The Works of John Ruskin*, Library Edition (London: George Allen, 1903-1912), 7: 205-6. In our own time, one is put in mind of Gregory Bohm speaking of the 'shape' (form) not the 'strength' (force) of the signal as what matters in communication, and Rupert Sheldrake's notion of "morphic resonance" in, e.g., *The Rebirth of Nature — The Greening of Science and God* (London: Rider, 1990), 88-90.

44 Cf. R. Panikkar, "Words & Terms," in M. Olivetti, ed., *Esistenza, Mito, Ermeneutica — Scritti per Enrico Castelli* (Roma: Archivio di Studi Filosofici, 1981), II: 117-133.

45 Signs are arbitrary epistemic pointers (from the Latin *signum*, a military banner); symbols are 'always more' than just pointers. (Cf. R. Panikkar's symbolic difference in *Worship & Secular Man* (New York: Orbis and London: Darton, Longman & Todd, 1973), 20-1, and "Words and Terms," cited above). Among Panikkar's points here and elsewhere: Terms are arbitrary signs, words (as living symbols) are not. H_2O is less than water; to be useful as a precise scientific term, H_2O must excise the breadth and variety of meanings in the multiplex symbol water. Further, any statement entails a truth claim. (NB: This statement cannot be refuted, since any refutation would have to claim some validity, some truth...). Consider the implications: If the statement "Language is arbitrary" is true, then its opposite — "Language is NOT arbitrary" — would have an equal

claim to truth. In such a scheme, there is no way to determine the relative value of two arbitrary but opposite propositions. Therefore power intervenes, power 'arbitrates' which 'entirely arbitrary' proposition is useful, or meaningful, or maybe just titillating for the moment. Only if language is *not* entirely arbitrary can the statement "Language is arbitrary" claim any degree of validity (in the restricted sense, for instance, that the sounds for words or the letters of the phonetic alphabet are arbitrary signs). Otherwise, any statement can arbitrarily be replaced by its opposite should the configurations of power change. Not so long ago, for example, we heard media commentators asking "When is sex not sex?," in the case of Bill Clinton's Oval Office shenanigans, or "When is a bribe not a bribe?" in the Salt Lake City Olympic scandal. Ezra Pound, by contrast, posted the Confucian slogan "Honesty is the treasure of states" in Rapallo, recalling Confucius' response when asked to what he would first set his mind if he were made ruler, "To call people and things by their names, that is by the correct denominations, to see that the terminology was exact." (GK 16).

46   Ralph Waldo Emerson, *The Works of Ralph Waldo Emerson* op. cit., "Shakespeare; or the Poet" (1850), Vol 2: 246.

47   Cf. Herbert Fingarette, *Confucius — The Secular as Sacred* (New York: Harper & Row, 1972).

48   One might adduce here Claude Levi-Strauss's distinction between 'game' and 'ritual' in the introduction to *The Savage Mind* (Chicago: University of Chicago Press, 1966). A game starts out in equilibrium — all players equally matched, following the same rules — but ends in disequilibrium, winners and losers. A ritual may have people starting out from very different points, and following very different paths, but the ritual is complete only when all parties are in harmonious equilibrium. The Confucian social order is conceived as a ritual; modern Western society, by contrast, is seen as a game, with its inevitable winners and losers. Cf. David L. Hall and Roger T. Ames, *Thinking Through Confucius* (Albany: SUNY, 1987), the most challenging and incisive recent effort to put aside the Western conceptual biases which have all too often obscured the aesthetic basis of Confucian ritual observance.

49   Cf. J. Needham, *Science & Civilisation in China* (Cambridge: Cambridge UP, 1956), cited by F. Capra, *The Tao of Physics* (Boulder: Shambhala, 1976), 280.

50   Fingarette, *Confucius*, 7-10. A little further on, he notes: "From this standpoint, it is easy to see that not only motor skills must be learned but also correct use of language. For correct use of language is not merely a useful adjunct, it is of the essence of executing the ceremony. ...From this perspective we see that the famous Confucian doctrine of *cheng ming* — the 'rectification of terms' or 'correct use of terminology' — is not merely an erroneous belief in word magic or a pedantic elaboration of Confucius' own concern with teaching tradition. ...Once we are aware of the ceremonial or performative kinds of functions of language, the original texts begin to read differently." (15).

51   ibid., 17.

52   Cf. my *Paradise & Ezra Pound, The Poet as Shaman* (New York/London: UP of America, 1983), Part II, "The Word as Image," and Illustrations, "The Image as Word."

53   Cf. R. Panikkar, *The Vedic Experience* (Berkeley/Los Angeles: University of California Press, 1977), Part I, B "The Word (*Vac*)," 88-112, e.g., from the *Tandya Maha Brahmana*, XX, 14, 2: "This (in the beginning) was the only Lord of the Universe. His word was with him. The Word was his second. He contemplated. He said, "I will deliver this Word so that she will produce and bring into being all this world." (107) Cf. also Panikkar's translation of St. Paul's Letter to the Hebrews, 11:3: "By faith we understand that the secular worlds have been formed by the speech of God, so that from the invisible the visible came to be."

54   Norman O. Brown, *Closing Time* (New York: Vintage/Random House, 1973), 104. Curiously John Ruskin, too, came in later life to put great store in excavating the "fossil poetry" in etymology, apparently turned in this direction by the work of his colleague at Oxford, indologist Max Müller.

55   William Shakespeare, *As You Like It*, II, 2, 16, W.A. Nelson and J. Hill, *The Complete Poems and Plays of William Shakespeare*, New Cambridge Edition (New York: Houghton Mifflin, 1942), 220.

56   Cf., again, R. Kerns, *Orientalism, Modernism & The American Poem*, especially the final intriguing chapter on Gary Snyder's poetic.

57   William Carlos Williams, *Paterson* (New York: New Directions, 1963), V: 207. For explication, see Scott Eastham, "by defective means — "Poetic Diction and Divine Apparition in William Carlos Williams' Later Poetry," *Sagetrieb* 5. 1 (Spring 1986) 17-27; and further analysis of this process in Scott Eastham, "Hear/Say: Ezra Pound and the Ten Voices of Tradition," *Rendezvous* 22. 1 (Fall 1987), 8-25.

58   Hans-Georg Gadamer, *Truth and Method*, op.cit., 432.

59   Cf. Hans-Georg Gadamer, "Text Matters," in Richard Kearney, ed., *States of Mind — Dialogues with contemporary thinkers* (New York: New York UP, 1995), 262-289.

60   Norbert Weiner, *The Human Use of Human Beings — Cybernetics and Society* (London: Eyre & Spottiswoode, 1950), Ch. VIII, 123-143, which, beyond mocking the mania for secrecy surrounding atomic research ("This demand for secrecy is scarcely more than the wish of a sick society not to learn of the progress of its own disease."), also reveals the limitation of his 'humanistic' approach to the language of nature as a 'code'. After discussing code-breaking in general, and praising the work of the epigraphers decoding the Rosetta Stone, Weiner says: "There is, however, one act of decoding which is greater still. This greatest single example of the art of decoding is the decoding of the secrets of nature itself and is the province of the scientist... Scientific discovery consists in the interpretation for our own convenience of a system of existence which has been made with no eye to our convenience at all. The result is that the last thing in the world to be protected by secrecy and (an) elaborate code system is a law of nature. ...It is perhaps impossible to devise any secondary code as hard to break as the natural code of the atomic nucleus." (137). — Unless of course it be the 'code' locked into the nucleus of the living cell. Weiner's understanding of what the scientist/humanist is up to governed quite directly Watson and Crick's labelling of DNA as a 'code'. Yet behind the code, Weiner himself concedes, lurks the notion that it is actually language... "In the problem of decoding, the most important information which we can possess is the knowledge that the message

which we are reading is not gibberish." (ibid.) Indeed, "in the interpretation... of (our) system of existence," most of the world's religions and philosophies had early on come to the realisation that it "is not gibberish."

61 Gregory Bateson, *Mind & Nature — A Necessary Unity* (New York: Dutton, 1979), 'Introduction', 17: "The shapes of animals and plants are transforms of messages. Language itself is a form of communication. The structure of the input must somehow be reflected as structure in the output. Anatomy *must* contain an analogue of grammar because all anatomy is a transform of message material, which must be contextually shaped. And finally, *contextual shaping* is only another term for grammar."

62 Cf., e.g., John Ralston Saul, *Voltaire's Bastards* (Toronto: Penguin, 1992), "The Rational Courtesan," 77-107; and *The Unconscious Civilisation* (Toronto: Penguin, 1997), "From Managers and Speculators to Growth," 117-157.

63 Cf. Jeremy Rifkin, *The Biotech Century*, op.cit., Ch 7, "Reinventing Nature," 197-226: "In the sense of the true meaning of art, then, genetic engineering is a rank deception. ...Making decisions over what genes to insert, recombine, or delete in an effort to 'alter', 'transform', and 'redesign' oneself and one's progeny is less an artistic expression and more a technological prescription. It is not art, but artifice. What some social theorists call the "Creative Age" is really the age of unlimited consumer choice. Unfortunately, we increasingly confuse the ability to choose with the ability to create, especially with regard to the new biotechnologies. Now that we can begin re-engineering ourselves, we mistakenly think of the new technological manipulation as a creative act, when in reality it is merely a set of choices purchased in the marketplace." (226).

64 In the midst of spinning one of his (very Poundian) tall tales about the conspiratorial domination of world trade by global pirates (like the East India Company), Buckminster Fuller inserts an observation of Pound's — presumably obtained directly, either in conversation or correspondence — pertinent to our presentation: "...when the Phoenicians began trading with people of so many different languages that, in need of a means of recording the different word sounds made by people around the world, the Phoenicians

invented phonetic spelling — Phoenician spelling — which pronounced each successive sound separately and invented letter
symbols for each sound. With phonetic spelling human written
communication changed very much — from the visual-metaphor-
concept writing of the Orient, accomplished with complex idea-
graphics (ideographs), several of which frequently experienced,
generalised cartoons told the whole story visually. It was a big change
from ideographs to the Phoenicians' phonetic spelling, wherein each
letter is a single sound — having no meaning in itself — and whereby
it took several sounds to make a whole word and many such words
to make any sense — i.e., a sentence. This is the historical event
that Ezra Pound says coincides with the story of the Tower of Babel.
Pound says that humanity (human language?) was split into a babble
of individually meaningless sounds while losing the conceptual
symbols of whole ideas."
R. Buckminster Fuller, "Legally Piggily," *Critical Path*, (New York:
St. Martin's Press, 1981), 75. Fuller also notes that invention of
the alphabet coincided with the Phoenician innovation of metal
money, replacing "cattle, lambs, goats, pigs — live money that was
real life-support wealth, wealth you could actually eat," but portable
and much handier for global pirates exploiting lines-of-supply. And
since there is today big money to be made from artificial plants and
animals, bio-engineers working for transnational conglomerates wish
us to believe that deployment of these new technologies is inevitable.

65 "Sacred secularity" is a phrase coined by R. Panikkar which
appears, e.g., in *The Cosmotheandric Experience — Emerging
Religious Consciousness*, S. Eastham, ed. (New York: Orbis, 1993),
121.

66 Cf. Mircea Eliade, *Shamanism*, Princeton (Bollingen) 1972,
Chapter VIII, "Shamanism & Cosmology"; and Lawrence E.
Sullivan, *Icanchu's Drum*, (New York: Macmillan. 1988), e.g.,
"Since the primordial sky is so often the object of the first real
separation, it betokens the very possibility of distance between one
kind of being and another; its continued transcendence guarantees
the symbolic life it signifies. ...Fear that the sky might fall becomes
more plausible when it is realised that only symbolic measures, the
elementary images of space and time, mark off the distance of heaven

from earth. ...The great cosmogonic preoccupation with raising the sky, or 'roof' of the first living space, emphasises that only by maintaining transcendence can the religious imagination uphold the distance between kinds of being. As a mythic experience and as a concept, transcendence makes symbolic life possible." (45-6).

67 Hugh Kenner, *Mazes*, (San Francisco: North Point Press, 1989), "The Making of the Modernist Canon," 42.

68 See R. Panikkar, *The Cosmotheandric Experience*, op.cit., Part One, "Colligite Fragmenta — For an Integration of Reality," 1-78. "The cosmotheandric principle could be formulated by saying that the divine, the human and the earthly — however we may prefer to call them — are the three irreducible dimensions which constitute the real, i.e., any reality inasmuch as it is real. ...What this intuition emphasises is that the three dimensions of reality are neither three modes of a monolithic undifferentiated reality, nor are they three elements of a pluralistic system. There is rather one, though intrinsically threefold, relation which expresses the ultimate constitution of reality. Everything that exists, any real being, presents this triune constitution expressed in three dimensions. I am not only saying that everything is directly or indirectly related to everything else, the *pratityasamutpada* of the Buddhist tradition. I am also stressing that this relationship... flashes forth, ever new and vital, in every spark of the real." (60).

69 Report of the Royal Commission on Genetic Modification available at http://www.gmcommission.govt.nz/.

70 Nicky Hager, *Seeds of Distrust — The Story of a GE Cover-Up* (Wellington: Craig Potton, 2002).

71 *Seeds of Doubt* (Bristol: Soil Association, 2002), which concluded that planting of GM corn, soybeans and oilseed rape since 1999 has already cost the US economy $25 billion, a disaster even in strictly economic terms. (www.soilassociation.org)

72 R. Panikkar. "The Myth of Pluralism: The Tower of Babel — A Meditation on Non-Violence," in *Invisible Harmony. Essays on Contemplation & Responsibility*, H.J. Cargas, ed. (Minneapolis: Fortress, 1995), 56-7. (Lines cited out of sequence).

73 Sharon Beder, *Global Spin — The Corporate Assault on Environmentalism* (Devon: Green Books, 1997).

74 Jacques Ellul, *The Technological Bluff*, G.W. Bromiley, trans. (Grand Rapids: Eerdmans, 1990), 187.

75 ibid., 145.

76 J. Ellul, *Propaganda — The Formation of Men's Attitudes* (New York: Knopf, 1965), 6.

77 Despite hundreds of labs and thousands of researchers around the world focused on precisely this problem, not to mention all the supposed 'breakthroughs' with which we are bombarded on the nightly news (publicity garners funding), nobody has yet come up with a safe, effective and lasting way to correct genetic errors in humans.

78 Cf. Cicero: a*nimi assidua et vehemens ad aliquam magna cum voluntate occupatio*, translated by Raimon Panikkar in his Foreword to S. Eastham, *Nucleus — Reconnecting Science & Religion in the Nuclear Age* (Santa Fe: Bear & Co., 1987), xxxii-xxxiii.

79 Juvenal, *Quis custodiet ipsos custodes*, literally, "Who is to guard the guardians themselves?", in *Satires*, 6, 1, 347.

80 Gregory Bateson, interviewed by Daniel Goleman in *Psychology Today*, May 1978, 44.

81 A 'controlled experiment' leaves one population of test subjects unaffected so as to check the effect of whatever is being tested. (By contrast, a 'control experiment' maintains identical conditions but alters a single variable, from which a causal factor may be inferred).

82 John Ruskin, *Unto This Last* (London, 1860), in *The Genius of John Ruskin — Selections from his Writings*, John D. Rosenberg, ed. (London: Allen & Unwin, 1963), 270.

83 Erwin Chargaff, "On the Dangers of Genetic Meddling," *Science* (192: 938-40).

84 William Shakespeare, *King Lear*, IV, vi. 55.

85 Wendell Berry, *Life is a Miracle — An Essay Against Modern Superstition* (Washington, DC: Counterpoint, 2000), p. 10.

86 ibid, 73-6.

Ezra Pound Works Cited in Text

(Number/Page) — *The Cantos*, New York (New Directions) 1970 (First printing of Cantos 1-117 in one volume).

Con — *Confucius — The Unwobbling Pivot, The Great Digest, The Analects*, New York (New Directions) 1969.

Con Odes — *The Confucian Odes — The Classic Anthology Defined by Confucius*, Cambridge, MA (Harvard University) 1954; New York (New Directions) 1959.

GK — *Guide to Kulchur*, London (Faber) 1938; Norfolk, Conn. (New Directions) 1938, 1952.

LE — *The Literary Essays of Ezra Pound*, London (Faber) 1954; Norfolk, CT (New Directions) 1954.

P — Personae — *The Collected Shorter Poems of Ezra Pound*, New York (New Directions) 1971.

SP — *Selected Prose — 1909-1965*, William Cookson, Ed., New York (New Directions) 1973.

T — Ezra Pound: *Translations*, New York (New Directions) 1963.

Illustration from Pound's *Confucius* p. 114 © Mary de Rachewiltz and Omar S. Pound (Courtesy Peggy L. Fox, New Directions)